Anatomy Without a Scalpel

Lon Kilgore

ISBN-10 0615390722
ISBN-13 9780615390727

Printed in the United States of America

Killustrated Books - Iowa Park - Texas

PREFACE

Anatomy is a daunting subject to everyone. Lots of things to learn. Lots of things to memorize. Lots of books, images, and charts to peruse. Is it worth the effort? Yes. Understanding how something is built helps us understand how it works. There is a biological and engineering adage - "form equals function" - that is quite important in exercise and sport. When we want to get better at an exercise we need to understand how the structure of the body contributes to its performance. Anatomical technical errors, something as small as too narrow of a bench press grip or running with the elbows held away from the body can significantly reduce performance. Simple changes in structure can also change function; adding mass to a muscle makes the angle of pull closer to the optimal ninety degrees for force application and losing mass reduces the number of muscle protein interactions that can produce force, the good and the bad. Anatomical applications abound everywhere in every exercise, every sport, and every body. As a trainee, especially if training on ones own without a coach or trainer, ignorance of the body's structure keeps us from performing at our best. As a trainer or coach, ignorance of anatomy prevents us from effectively and reliably teaching others to train to reach their goals ... that directly affects our professional success. There are examples of coaches, trainers, and trainees who have succeeded without understanding science, but they did so by chance or by mimicry. They were equally likely to fail. This book is a raw product, a simple presentation of my lectures, lecture notes, sketches, and essays that have evolved over the years - hopefully in an easy to read format. It is intended to tilt the odds in your favor and set the anatomical foundation for the teaching and programming of exercise.

Lon

THANKS

Thanks to Alissa Donaldson, Steve Hilton, Josh Wells, Bethany Talley, Shelly Hancock, Lauren Frech, Justin Lascek, Gant Grimes, Bryan Fox, Dr. Kathy Weezner, and Officer Alan for consenting to be photographic subjects. I apologize for the interruptions in your workouts caused by having a little maniac pull you out of a perfectly good training session for a photo or twenty. Your contribution and patience is much appreciated.

Thanks to my many students who facilitated MY learning and the CrossFit Journal who published excerpts of this book and gave me a means to gather valuable objective feedback about its contents. Thanks to Carrie, Mary, and Mike for their feedback and to Greg for giving me a forum that actually gets read.

CONTENTS

"Self-education is, I firmly believe, the only kind of education there is."

- Isaac Asimov

"The brain is like a muscle. When it is in use we feel very good."

- Carl Sagan

1 – THE MEASURE OF A MAN

Simple questions are sometimes the most profound. And answers to simple questions about exercise sometimes do not exist in print; rather, they are often intuitive to skilled coaches or contained in knowledge that is passed on in the lore of the gym rather than recorded in books or formal training programs. Expert coaches, trainers, clinicians, and professors sometimes take it for granted that what is basic to them is likewise simple, common knowledge to all. During the course of shooting video footage for a DVD project spearheaded by barbell coaching maestro Mark Rippetoe, Rip made a number of comments about arms being short, legs being long, and various and sundry other references to body segments not being of the usual proportions. One particularly humorous comparison of one of our acquaintances to a Tyrannosaurus Rex made Katie, a trainer from northern California, ask a question: "How do you know someone's arms are longer or shorter than normal?" It was a simple but very good, insightful, and germane question. But it is a question that, as far as I know, is not treated in the exercise literature anywhere.

Movement, specifically technical movement in exercise and sport, is subject to anthropometric and geometric influences. This means that how people's bodies are put together and the relative sizes of the various parts affect how they look and perform when doing certain movements. Just think about obvious cases of this truth – NBA centers and NFL offensive linemen, for example. Their build suits the demands of their sport and position, and so the best players in a given physical sport usually have similar dimensions. Soviet sports scientists even had a set of target anatomical dimensions they used in selecting developmental athletes in various fields to increase the likelihood of individual and team success. Championship teams are frequently built by recruiting players with the right bodies and skills – as much as by elite coaching.

The average trainer, coach, or physical educator must have a functional understanding of how differing anatomical phenotypes (different body dimensions and body-segment lengths) affects the way proper technique looks. To do this, one must first have a reference point and a means to mentally and visually comprehend typical segmental relationships. Some of this is intuitive in good coaches, but, as Katie's question makes clear, we also need a more concrete way to determine whether an individual's torso is longer or shorter than average for someone of that height, whether the arms are of average length or not, or whether there is a difference in leg length that is significant to the movement in question. If there is a difference it needs to be localized. Is it in the upper arm vs. forearm, or in the shin vs. the thigh? A savvy coach will have a knack for this determination. It is a handy skill. Being able to see, at a glance, how a trainee's body dimensions compare to an average template helps us place the trainee in correct, efficient, and safe exercise positions. We cannot teach our trainees how to exercise to their best benefit for the biggest gain in fitness if we are oblivious to anthropometric considerations. Indeed, we may, without intent, place trainees in positions that can decrease their efficiency and even increase their risk of injury.

VITRUVIAN MAN & THE MEASURE OF BODY SEGMENTS

Fortunately, we do not need to develop a new and elaborate system for doing such an analysis. We can simply take a trip back to the Renaissance to revisit the works of Leonardo da Vinci. Virtually everyone exposed to exercise, anatomy, sport, or da Vinci has seen his drawing of a man in a circle in a square, known as the Vitruvian Man (after Vitruvius, an architect contemporary of da Vinci, who analyzed human dimensions based on four finger widths). Lots of sport and exercise academic programs and even fitness businesses use the Vitruvian Man in their logos. But the Vitruvian Man is not just a cool drawing; it is da Vinci's attempt to map average human dimensions (the average human phenotype or average anthropometry). Da Vinci's notes on human proportions include the following observations:

- The length of a man's outspread arms is equal to his height.
- From the bottom of the chin to the top of the head is one eighth of his height.
- The greatest width of the shoulders contains in itself the fourth part of man.
- From the elbow to the tip of the hand will be the fifth part of a man.
- From the elbow to the angle of the armpit will be the eighth part of man.

This may be more precise than we need for determining proper movement positions, but what is interesting is that for the past half millennium or so, while physical educators, coaches, and other fitness professionals have ignored it, artists and art teachers have derived systems of representing the "average" human form largely based on Da Vinci's works. In these systems the human head is used as a reference length in determining the total length of the body, the length of body segments, and their placement on the body. The most common models use either seven and a half or eight head lengths to establish the height of the average body (figure 1-1). This is not an absolute model, rather a general guideline for convenient reference for beginning fitness professionals. You can practice eyeballing anthropometry on the web by searching out full-length photographs of your favorite athletes (or just normal people) and using a clear ruler to measure head lengths. You will find variation. You might find that Flozell Adams is 8.3 heads high but Brett Favre is 7.6, Brock Lesnar is 7.4 and Randy Couture is 6.9, or that Kerri Walsh is 8.1 and Gina Carano is 7.2. There are even some occupations that seem to select for certain ratios, as a random assessment of the Victoria's Secret catalog revealed a range in head lengths from 8.2 to 8.5. As the "coaching eye" develops from years of observation and analysis of the human form, less reliance on artificial constructs will be needed.

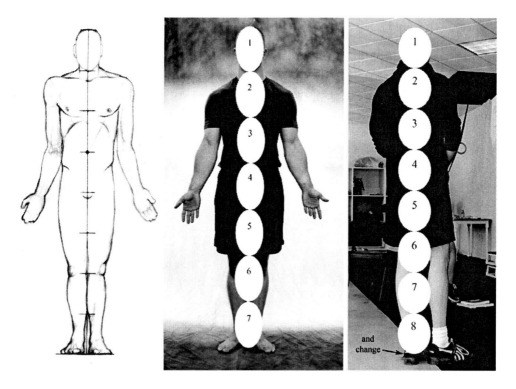

Figure 1-1. The 8-head model of representing average human dimensions. The average human you see in the gym and on the street will vary from the Vitruvian standard (left). You will see some trainees with a full head length less than eight and some with a little more than eight.

ANATOMICAL LANDMARKS

In these models, various anatomical landmarks are a specific number of head lengths distant from the top of the head or the bottom of the feet (figure 1-2*)*. A perceptive coach can use this generalized system to determine whether a trainee has normal anthropometry or has particular segments that are longer or shorter than predicted by the "average" model. We can figure out whether a trainee's upper arm is long or short if we know that, on average in the eight-head model, the elbow is at about the same level as the belly button. An elbow observed to be well below the level of the belly button indicates an arm—or at least an upper arm—longer than average, given an average torso length. Likewise, an elbow well above that level indicates either a shorter than average segment (arm or upper arm) or a longer than average torso (figure 1-3). Deviations from average anthropometry necessarily change the geometry of positions and movement.

3

This means that the same exercise movement looks different, sometimes subtly and sometimes dramatically, in individuals with different segment lengths.

You will see these types of variation in anatomical structure regularly if you take the time to observe carefully. One of our goals in this book is that you will be able to teach individuals just like these how to exercise correctly and safely by accommodating their structural differences. Sometimes segment length variations are not apparent until you see someone moving, especially when the trainee is exercising in loose clothing.

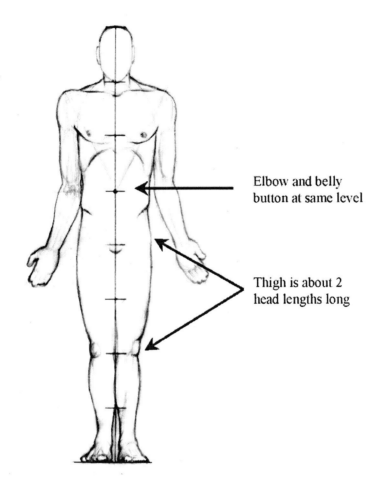

Elbow and belly button at same level

Thigh is about 2 head lengths long

Figure 1-2. Some basic landmarks and features of the 8-head model.

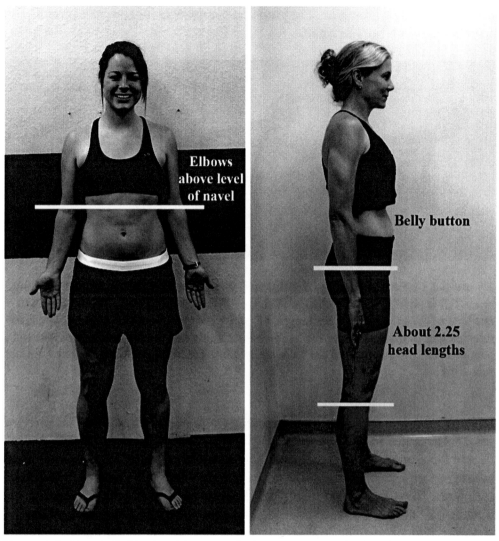

Figure 1-3. Two individuals with anatomical segments that deviate from average. On the left, note that the elbows do not extend to the level of the navel or below and that the fingertips are at the level of the hip joint rather than at the level of the upper thigh. On the right, note that the length of the legs relative to the torso is not average.

ANATOMICAL SEGMENTAL DEVIATIONS & EXERCISE

In barbell exercise, correct and efficient positioning means the difference between a safely executed exercise yielding big gains and an unsafely executed exercise yielding small increases in strength with the potential for chronic soreness. An example of this is in the starting position for the deadlift or power clean. As shown in figure 1-4, something seemingly innocuous, like arm length,

5

can make a correct starting position look quite different in individuals with different arm measurements. The situation becomes more complex if multiple segments deviate from average.

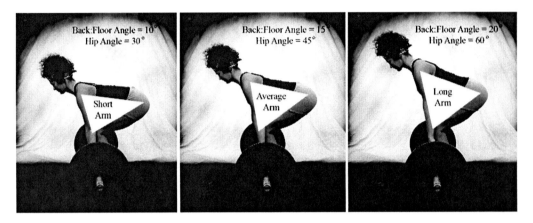

Figure 1-4. Comparisons of the deadlift start position with short vs. average vs. long arms (left to right). Note the change in shape of the central triangle. As the length of one side of the triangle changes length, all of the component angles change. The most relevant angles to observe are the back-to-floor angle and the hip angle (femur-vertebral column). This makes each of the technically correct starting positions depicted above look different.

In some instances the anatomical variations may be so extreme that an exercise must be modified in order to accomplish its purpose with adequate safety. Figure 1-5 demonstrates such a case. The coach is presented with an individual whose anatomical structural differences require wholesale changes in exercise technique. In setting up a correct deadlift for an individual with average dimensions, the bar is placed over the mid-foot and under the middle of the shoulder blade. However, in this case, the model has very long thighs, making the correct position for a person of average dimensions impossible. If standard technique were to be insisted upon with this individual, the position of her hips, shins and the bar relative to the shoulders would not be optimal or even safe. When the hips are lowered, the shins will move forward, the bar will be pushed well out over the toes, and the shoulders will be behind the bar. The solution was to prescribe the Sumo style of deadlift thus allowing the mid-foot, bar, and shoulder blade to come into proper alignment. While not optimal form for comprehensive development, it is a solution that allows the inclusion of the deadlift in this person's training in as safe and as effective manner as possible.

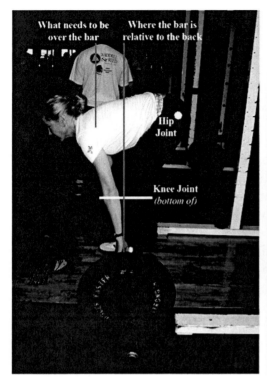

Figure 1-5. In the picture on the left, the individual is in an unsafe position for the deadlift because of very very long thighs. If the hips are lowered the shins push the bar forward over the toes. On the right, the Sumo style deadlift places the mid-foot, bar, and shoulder blades in proper alignment for a safe deadlift. This adjustment would not be an effective adaptation on the clean as it would add problems with the top portion of the pull and catching the weight on the shoulders.

When faced with extreme deviations, thoughtful analysis and experimentation is required in the modification of the exercise. Do not force the individual to attempt to occupy spaces and positions that their bodies cannot geometrically assume (this is not in reference to flexibility issues).

In regards to spatial orientation of muscles, bones, and joints, it is important for a trainer and a trainee to understand that how a person is built directly affects how an exercise should be done. Proper technique provides for efficient and unrestricted movement. Improper technique reduces efficiency, restricts range of motion, and ultimately can increase the likelihood of injury. As illustrated in figures 1-4 and 1-5, segmental length differences alter the joint angles and subtly (and not so subtly) affect how proper technique looks. There are also instances, common to all trainees where anatomical relationships dictate proper technique. An excellent example of this can be found in determining appropriate stance and foot placement during the execution of the back squat.

Frequently in trainees improperly coached in the squat, a bottom position is not achieved without rounding the lower back. Walk into virtually any weight room in the country and you will see examples of this. These individuals are frequently told that they have poor flexibility and end up spending an inordinate amount of time stretching their hamstrings – usually to no avail. What is actually occurring here is that the trainee is not being coached to assume a correct stance that allows the ilium (one of the hip bones) to drop in between the femurs (thigh bones). If the stance is too narrow or the knees are pointed directly forward, the ilium restricts the femur's range of motion and prohibits a proper squat bottom position (figure 1-6). In chemistry we would call this steric hindrance. This simple example is not an isolated one, every exercise involving moving parts will be affected by how the body is built. An effective coach or trainer has to be aware of these relationships in order to teach proper technique.

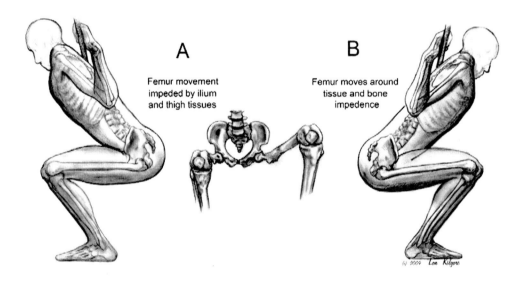

Figure 1-6. In panel A, the femur is aligned improperly underneath the anterior iliac crest and will impede the excursion of the hip, preventing the hip from reaching a safe and effective bottom position. Panel B shows a wider stance and more angled foot placement that allows for the two bones to pass unrestricted and allows the trainee to reach a correct bottom position.

ACCOMMODATING SEGMENTAL DIFFERENCES

Recognizing segmental differences is half the battle; knowing what to do with them is the other half. Segmental differences can affect grip, stance, start position, jump position, and pretty much every aspect of barbell exercises. For example, in the snatch we want a grip as wide as feasible to reduce the distance the bar needs to travel; short arms require a narrower grip and long arms require a wider grip on the bar. Similarly:

- Long forearms usually require the trainee to assume a wider grip when racking the weight on the shoulders for pressing or when receiving the bar during a clean.
- Long thighs necessitate a higher hip starting position (and therefore a more horizontal back position) off the floor in pulling motions.
- Short thighs mean a lower hip position and a steeper back angle.
- Long arms mean that in the power clean, the bar will be lower on the thighs in the jump position than we would like, but things are the way they are.
- Any time a guy with a big beer gut pulls a bar off the floor, the larger-than-average belly segment must change the geometry of the stance. A little wider stance with the toes pointed out a little more than normal is needed to allow the proper lifting mechanics to happen, and to give the belly a little room to hang out.
- If the trainee is built like Magilla Gorilla and the bar is only two inches above the knee when it's directly under the most forward point of the shoulder blade, reinforce that to the trainee as the correct position based on his physical construction. Don't tell him to hit the bar higher on his thigh; that will likely cause him to start pulling with bent arms.

SOLVING MOVEMENT PROBLEMS

Anyone who lifts or works with lifters, whether as a professional trainer, an amateur coach, or just a lifting buddy needs to recognize that there isn't a one-size-fits-all template for body angles that every body must conform to. If you recognize an anatomical deviation and don't know how to approach it, think about it, draw it out, experiment, and talk to other trainers. Although solving this kind of movement problem is a complex geometry problem, with a little time and creative mental effort, you can typically arrive at a solution for your trainee.

Anatomical segmental differences can also affect a multitude of other exercise or sport skills. Just think of every televised boxing and mixed martial arts contest you have ever seen. Remember the "tale of the tape" comparing the two combatants? Does arm length affect the fighting style used by a combatant? Of course it does. Look at the oarlocks on a rowing shell; why do they have the capacity for adjustment? Although elite rowers are generally tall, there is enough segmental variation among them that equipment settings need to be adjustable to maximize performance.

In summary, the three important things in solving movement problems are:

- Ability to identify anatomical segments and assess their length relative to average
- Knowledge of the key points of effective exercise positions
- Understanding of how atypical anatomical segment lengths affect these movement patterns

It is readily apparent that at this, the most simplistic level of movement analysis, a *detailed* knowledge of anatomy is not required. However, even at the most basic level, a coach or trainer in any physical discipline must develop a working knowledge of anatomy just to be able to teach and coach movements correctly (an understanding of very basic geometry helps a bit too). Anatomy is not just important to the biology and medical professions. Learning this oft-neglected coaching skill first marks the initial step in the application of anatomical principles to movement mastery.

2 - PLANE AND SIMPLE

Anatomy is an intimidating topic to many, but it is an important area of study relevant to a multitude of professions, including those associated with the teaching of exercise. In Chapter 1 we examined how the recognition of body segment lengths that deviated from average affected how a movement would look. It was our first step at developing an "eye" for coaching. In the next step, we will take a look at how to describe human movement in very specific spatial and directional terms and simplify them for easy communication in teaching.

ANATOMICAL POSITION

Let's start with the basic reference position used in anatomy – "anatomical position" (figure 2-1). Realize right off the bat that anatomical position is an artificial construct intended to provide anatomists a means of describing the positions of various anatomical features occurring on and in the human body. It was derived from the position that dissected bodies assumed as they lie on the dissecting slab – flat on the back with the palms of the hands facing up. Test this out by flopping flat on the floor yourself, relaxing completely, and seeing what your hands do. The palms will face upwards (mostly). But anatomical position is presented to us in the living body as a standing position. If I stand up in a relaxed position my arms will hang in a way in which my palms face the hip or slightly to the rear. Interesting eh?

Figure 2-1. Relaxed standing posture (left) compared to anatomical position (right).

The fact that anatomical position is not representative of normal posture is not too important – having a defined reference position is. By having one absolute reference point, anatomists worldwide view and describe a feature on or in the body in a similar manner, avoiding confusion.

Think of anatomical position as magnetic north on your compass. A compass allows people from all around the world to be dropped into unfamiliar terrain and find their way to a specific destination. Likewise, when we need to describe the body's precise anatomical locations and directional movements, the terminology described for anatomical position serves us well.

ANATOMICAL TERMS YOU NEED TO KNOW

Orienting yourself to the body is important whether it is at a dissecting table, standing up, or in a convoluted exercise position. Every exercise professional needs to know where anatomical parts are relative to each other, the earth, and to any implement used during the exercise. Anatomists have a nice set of uniformly understood terms describing just that. These will be necessary for the exercise professional to learn. They are arranged in opposing functional pairs as follows:

Front - Back

Anterior - A structure is anterior if it lies in front of another structure (figure 2-2). For example, the toes are anterior to the heels.

Posterior - A structure is posterior if it is located behind another structure (figure 2-2). The erector spinae, the long muscle group running vertically along the back, is posterior to the abdominal cavity.

Figure 2-2. Anterior or front view of the upper body (left) and posterior or back view of the upper body (right).

Near - Far

Proximal - This term is usually associated with the extremities but is relevant to all structures. It describes a structure as being closer to the center of the body or to the beginning of the extremity than another structure (figure 2-3). For example, the knee is proximal to the foot.

Distal – Meaning the opposite of proximal, "distal" describes a structure farther from center or from the beginning of the extremity than another structure (figure 2-3). For example, compared to the elbow, the hand is distal to the shoulder.

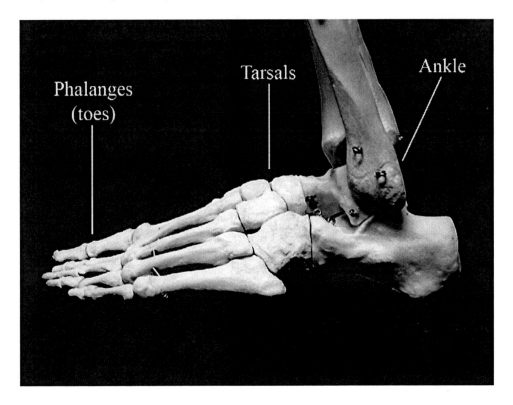

Figure 2-3. The tarsals are proximal to (close to) the ankle joint. The phalanges (toes) are distal to (away from) the ankle joint.

Top - Bottom

Superior – Use of this term signifies that a structure is higher than another structure (figure 4). The head is superior to the pelvis, for example.

Inferior - Designates that an anatomical structure is lower than another (figure 2-4). For example, the chin is inferior to the nose.

Figure 2-4. A superior structure is above another. In the top image, the skull is superior to the pelvis, or we could say that the pelvis is inferior to the skull (below the skull). We can move and change this orientation during exercise. In the image on bottom, the pelvis is superior to the skull.

Middle – Side or Inner - Outer

Medial - When a structure is closer to the cardinal sagittal plane (center) than another structure, it is medial (figure 2-5). An example would be that the sternum (breastbone) is medial to both shoulders; another would be that the spinal column is medial to the ribs.

Lateral - A structure that lies farther away from the cardinal sagittal plane than another structure is termed lateral (figure 2-5), for example, the shoulders are lateral to the sternum.

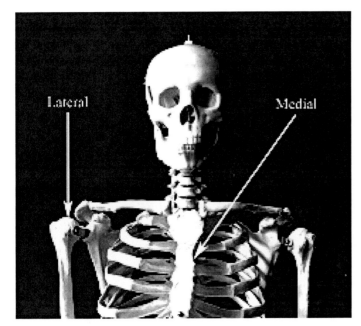

Figure 2-5. The sternum is medial, or in the middle of the body. The shoulder joints are lateral, or away from the midline.

THE CARDINAL PLANES

Now that we have a starting point we can take a few more steps toward describing human movement. Anatomists use the convention of anatomical planes in describing anatomical features and their locations.

The cardinal planes intersect at the body's theoretical center of gravity or center of mass. Christian Wilhelm Braune first reported this intersection in the late 1800's. Planar terminology was first used to describe the cuts made in the process of dissection.

The planes are also useful in that they allow one to describe the orientation of a feature relative to a plane. You might even be able to draw a loose parallel between using anatomical planes to describe the position of an anatomical feature and using latitudes and longitudes to describe the location of a geographic feature on the earth.

The anatomical planes have been hijacked and renamed by the exercise sciences. Every text on kinesiology or biomechanics has a section on the "Planes of Motion", which, quite simply, are the anatomical planes. On the other hand, most anatomy texts refer to the "anatomical planes" as such, but only in passing because they expect the reader to be familiar with the "cardinal" (important) planes – *frontal, transverse,* and *sagittal* (figure 2-6).

15

Frontal – The frontal plane divides the body into front and back halves as it passes side to side (shoulder to shoulder).

Transverse – The transverse plane divides the body into top and bottom as it passes perpendicular to the long axis of the body.

Sagittal – The sagittal plane divides the body into right-hand and left-hand sides as it passes front (toe side) to back (heel side).

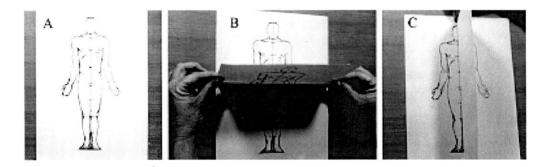

Figure 2-6. A simple exercise to demonstrate the cardinal planes. (A) The paper represents a frontal plane passing through the body. (B) The folded paper protruding from the frontal plane represents the transverse plane. (C) The folded paper protruding from the frontal plane represents the sagittal plane.

AXIS OF MOVEMENT

In exercise texts we see references to "Axes of Movement". If you remember your high school geometry, an axis is an imaginary line around which something rotates. Every plane of movement has an axis. To visualize the axis of a plane and the rotation around it, put a flat piece of paper on the table in front of you. Let's say the paper represents the frontal plane. Now draw a large human stick figure on it in anatomical position. When looking at the paper now, you are examining the representation of the human (the stick figure) in a frontal perspective. If you place your pen point on your stick figure's belly button and spin the paper, it is rotating around a *frontal axis*. The axis of a plane extends perpendicular to the plane. Now carefully fold that paper in half, the crease running left to right, hopefully at the point the legs of your stick figure begin. The crease represents an axis of rotation, in this case the rotation of the upper and lower body at the hip. If the top half of the paper is folded down, the upper body has rotated around a *sagittal axis*. If you fold the paper with the crease from top to bottom you have created a *transverse axis*. OK. That was fun. Let's move on to something a bit more useful. Really, when have you or any coach or trainer you know used these terms? It is good information needed to understand anatomy but not necessarily useful in communicating information to clients or athletes.

HOW TO DESCRIBE MOVEMENT

If planar descriptions are elemental to the study of anatomy but not so elemental in practical applications to describing movement, how does one describe movement?

There exists a standard set of directional terms that specifically deals with both locating anatomical features and describing their movements. Many of the terms are actually familiar. Having used the terms "flex" and "extend", you are already part way to a simple understanding of how to describe movement.

Bend - Straighten

Flexion – This movement occurs in the sagittal plane, anterior-posterior (figure 2-7a). A characteristic of flexion is that the angle formed by the joint decreases. For example, the hamstrings will flex the knee joint from 180° at full extension to around 60° at full flexion; the number gets smaller as the joint flexes. In the gym we frequently refer to flexion as "closing" the joint. Think of it like closing a book. As the book closes, the two covers go from 180° to 0°. It is worthwhile to note here that it is the joint "flexing", not the muscles. Muscles contract or relax to flex or extend joints. There are always exceptions to things. Tell someone to flex their arm. They will automatically bend their elbow thus moving the forearm. This is flexion of the elbow not flexion of the arm at the shoulder. We have to pay attention to what we say and how we say it. As we will see in a moment, anatomical terms can be easily misunderstood.

Extension – As in flexion, extension occurs within the sagittal plane and is anterior-posterior. In fact, it is the opposite of flexion (figure 2-7b). The angle formed by the joint increases throughout the movement. Extension is a very common coaching term in the gym and in sport. An analogy here is the opening of a book, increasing the angle between the two covers from 0° to 180°. For example, the hip angle opens as you rise from sitting to standing. Exceptions occur here, as well as opportunity for miscommunication. Usually when we tell some one to extend their arm, they lift it forward and upwards. This is actually flexion of the shoulder. Pushing the arm backward behind anatomical position is extension, not raising it overhead. Tricky. This is a definite example of where anatomical terminology can confuse. A coach or trainer's job is to cut through the confusion by giving clear instruction using a vocabulary the trainee understands.

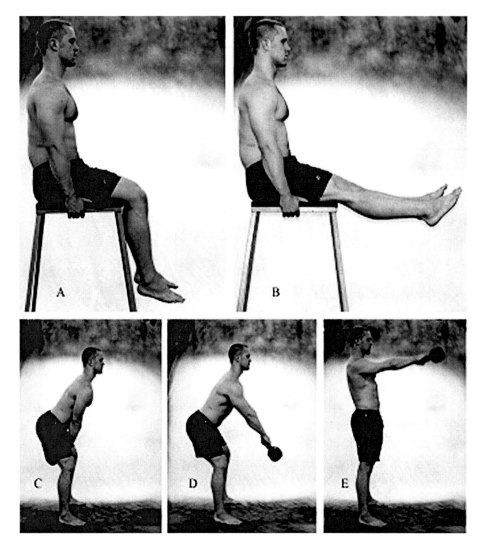

Figure 2-7. Flexion (a) and extension (b) of the knee joint. It is apparent from the photograph illustrating flexion that muscular involvement is not a pre-requisite of flexion or extension. Muscles do not flex, joints do. Flexion and extension in multiple joints is the norm in human movement. A simple task such as a kettlebell swing (C, D, and E) demonstrates simultaneous extension of the knee and hip along with shoulder flexion.

Move Away – Return

Abduction - Abduction occurs within the frontal plane and describes a movement taking a body part away from the mid-line of the body (figure 2-8a). A simple concept is that to abduct is to take away (i.e., a kidnapper takes someone away or abducts them).

Adduction - Like abduction, adduction occurs within the frontal plane, but is a movement of a body part towards the mid-line of the body (figure 2-8b).

Figure 2-8. (A) Abduction. (B) Adduction. These are pretty extreme examples of abduction and adduction of which most trainees are not capable. Some people would call this hyperabduction and hyperadduction. BUT "hyper" specifically refers to movement beyond the individual joint's range of motion capacity. Hyperextension, hyperabduction, hyperflexion, hyperadduction, hyper-anything generally represents an injury. Years of ballet training have made larger than normal range of motion abduction and adduction a normal and non-stressful movement for our model here.

Spin – Fling

Rotation – As its name implies, rotation is a pivoting motion around the long axis of the body or any body segment (figure 2-9).

Circumduction – This can occur within any plane and refers to a movement where the joint acts as a pivot point and the distal segment then moves in a circle around the joint (figure 2-10). In general, it is a combination of flexion, abduction, extension, and adduction ordered into a defined sequence.

Circumduction can only occur at any joint that is capable of movement in two or more planes. An extremely common example is an underhand softball pitch.

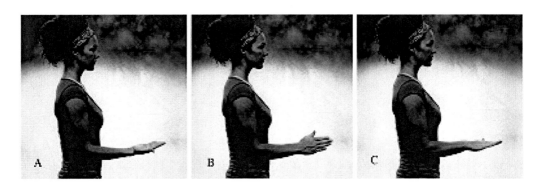

Figure 2-9. In the sequence above, the radius (thumb side of the hand) is rotated over the ulna. Another way to say this is to say that the right hand at the far left (A) is **supinated** and the right hand to the far right (C) is **pronated**. Note that in anatomical position the thumb is pointed to the outside also. This position can also be accomplished by movement of the shoulder and is called external rotation of the humerus. It is easy to discriminate between these to rotational movements – simply bend the elbow as above to demonstrate supination and pronation. Keep the elbow straight and the wrist locked to demonstrate internal (thumb to thigh) and external rotation.

Figure 2-10. Circumduction at the shoulder.

PRACTICAL APPLICATIONS

All of these terms – flex/extend, abduct/adduct, rotate, and to a lesser extent circumduct – can be used to teach a client or athlete correct form. There are lots of other specific terms like invert, evert, depress, elevate, and more that have specific meaning within specific anatomical segments (table 2-1) but the basic few described here will suffice for most exercise teaching purposes.

Movement	Body Parts Affected	Description
Flexion	Any applicable joint	Bending movement where the joint angle decreases
Extension	Any applicable joint	Straightening movement were the joint angle increases
Abduction	Any applicable joint	Movement of a segment away from midline
Adduction	Any applicable joint	Movement of a segment towards midline
Rotation	Any applicable joint	Circular motion around an axis
Elevation	Any applicable joint	Movement upwards
Depression	Any applicable joint	Movement downards
Internal rotation	Some appendicular skeletal joints	Rotation of the part anteriorly
External rotation	Some appendicular skeletal joints	Rotation of the part posteriorly
Circumduction	Ball and socket joints	Flextion, extension, abduction and rotation around the joint
Pronation	Elbow and wrist joints	Rotation of the hand to a palms down orientation
Supination	Elbow and wrist joints	Rotation of the hand to a palms up orientation
Dorsiflexion	Ankle	Lifting the toes and ball of the foot up
Plantarflexion	Ankle	Pushing the toes and ball of the foot down
Eversion	Ankle	Rolling the ankle with the sole of the foot facing out
Inversion	Ankle	Rolling the ankle with the sole of the foot facing the midline
Opposition	Thumb	Moving the thumb and fingers together
Reposition	Thumb	Moving the thumb away from the fingers
Protrusion	Jaw	Moving the jaw move forward (towards underbite)
Retrusion	Jaw	Moving the jaw backwards (towards overbite)
Protraction	Shoulder	Movement of the shoulders forward (abduction)
Retraction	Shoulder	Movement of the shoulders backward (adduction)

Table 2-1. Basic anatomical terms and their descriptions.

In fact, you will probably have to come up with fifteen different ways to tell a trainee to flex, extend, abduct, adduct, or rotate a joint or body part without using the correct anatomical term. Instead of saying "abduct your femurs" during a squat, "push your knees out" would get the message across. Instead of telling the trainee to "internally rotate the humerus" in the set-up for a clean, "point your elbows out" will elicit the response you want – correct engagement of the latissimus dorsi for proper force application. In this application of anatomy there are two tasks to be accomplished. They are:

First, the detection of segmental deviations from a reference standard (the prototypical example of good technique). This is accomplished through an understanding of the anatomical nature of the movement, what is actually occurring, and where it is occurring.

Second, an ability to convey anatomical feedback to your trainee in a vocabulary that is understandable to them.

The former point is science applied to exercise; the latter is, at its core, part of the art of coaching. Failure to accomplish either makes for a less successful coach or trainer.

"Anatomy is destiny."

- Sigmund Freud

"If anatomy is destiny then testosterone is doom."

- Al Goldstein

3 - GETTING SOME LEVERAGE

The human body is a complex system of levers that enables us to perform everything from very basic to extremely complex movements. Individually, levers are simple machines and in the body, muscles provide the force required to move by virtue of its being the force application component (motor) of a lever system. We can think of the body as a collection of machines. Movement machines are made up of several important components:

- *Bones* make up the rigid framework of the lever machine through which force is transferred.
- *Joints* provide the axes around which rotation occurs.
- *Muscles* generate the force (or tension) needed to cause overt movement.
- *Tendons* attach muscle to bone and as such are an intermediary, important and crucial to human movement. When a muscle produces force it is transmitted to the bone via tendon.

In the human body all three possible types of lever—first, second, and third class, are represented anatomically (this is arguable as one type requires special circumstances to be present; see Second Class Levers and figure 3-2).

CLASSIFYING LEVERS

When classifying levers the following entities serve to assess a lever system by type:

- Point of *force* application (F) – the site of muscle attachment and action
- Fulcrum or *axis* of rotation (A) – the joint to be flexed or extended
- Point of *resistance* application (R) – the site of the load to be moved

When these three entities, Force—Axis—Resistance, are laid out linearly, the space between the point of resistance application and the axis (fulcrum) is the *resistance arm*. The space between the point of force application and the axis is the *force arm*.

Understanding the relationship between force, axis, and resistance is at the root of ones ability to recognize and set up appropriate lever configurations and competently monitor technique. The squat, for example, is accomplished by a collection of levers. Something as simple as placing the bar high on the traps or low across the scapular spines alters moment arms and joint angles, changing the nature of the forces generated and experienced by the body, which in turn changes the nature of adaptation the exercise will produce, for the good or the bad. If we introduce inappropriate lever configurations or inadvertently create extra lever arms, we enable poor technique, poor force transference, and poor fitness results. Learn how to recognize levers in order to be able to reinforce good technique and eliminate the bad.

FIRST CLASS LEVERS

First class levers are the most efficient and probably the most identifiable (the teeter-totter, figure 3-1 A). In fact, the use of this type lever is extremely well documented throughout human history and spawned this quote by Archimedes in 220 B.C., "Give me a place to stand and a lever long enough and I will move the world". In this class of lever the fulcrum always lies between the force and the resistance (F-A-R).

A first class lever can perform all four basic machine functions;

- Balance forces
- Change direction of the force applied
- Modify (increase or amplify) the force applied
- Modify the speed and/or the range of motion

The first class lever has a mechanical advantage. Think of the teeter-totter, exactly 10 feet long end-to-end, the fulcrum at dead center, and a 50-pound kid on each end. What will happen if they just sit there? Right. Nothing. The forces are balanced and without a little leg push from one of the kids no one goes up, no one goes down. Now think about when you were a kid and how hard you had to push to go up. Not very hard. That's the lever system giving you a little advantage.

Putting this into an anatomical perspective, the same situation is present where the skull sits atop the cervical vertebrae (figure 3-1). How hard do you have to work to tilt your head from side to side or from front to back?

What happens if you move the teeter-totter's fulcrum to a point where one arm is 4 feet long and the other is 6 feet long? Which side has an advantage? Well, if you draw the short arm, you are only along for the ride. The moment around the axis of rotation (fulcrum) is a product of the force applied and the distance from the force (kid on the end) to the axis. Simple math tells us that the shorter arm is at a disadvantage. That means that the kid on the long arm can keep the kid on the short arm up in the air as long as he wants. It also means that to balance the force to where the asymmetric teeter-totter works as desired, a heavier kid needs to be on the short arm – or scoot forward.

Figure 3-1. An examples of a first class lever. The lever is causing a lateral tilt of the head (anterior view). Force = Neck muscle contraction, Axis = Atlas vertebrae, and Resistance = Weight of head. The size of the arrow indicates the relative magnitude of muscular force exerted.

SECOND CLASS LEVERS

In second class levers, the resistance is in the middle, between the axis and fulcrum (F-R-A). The most easily identified example of this class of lever is the wheelbarrow. In this arrangement the force arm is always longer than the resistance arm and this favors force production. According to Gray (of Gray's Anatomy – Myology Section discussion of levers), there are no second-class levers occurring in the body. In an unweighted human producing unresisted joint movement in the lab, this is true. However, in sport and exercise there are several examples of the body as a whole acting as a second-class lever. An easy example of this is the push-up (figure 3-2).

Figure 3-2. The push-up, a second class lever in exercise. The body is pushed up by the upper body musculature as it rotates around the ball of the foot on the floor. Force = arm muscle contraction; Resistance = body weight; Axis = ball of the foot.

25

THIRD CLASS LEVERS

The third class lever is the most common type of lever in the human body. The force applied is in the middle, between the resistance and the axis of rotation (R-F-A). In this lever arrangement the resistance arm is always longer than the force arm. The easiest and most recognizable such lever in the human may be the bicep and elbow (Figure 3-3).

Think of the resistance being in the hand and the elbow as the axis. The force applied to flex the elbow, although produced in the upper arm, is applied to the system between the elbow and the hand. This is a good reason to know where muscles attach at both ends, it will aid in understanding physically how they work.

There are also examples of human-implement interactions acting as third class levers; a baseball bat is a good example. The spatial arrangement of lever elements in this class results in a mechanical advantage biased towards increased range of motion and increased velocity of movement.

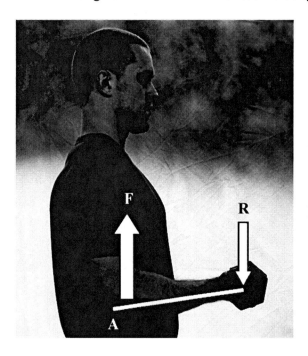

Figure 3-3. One of the exercises of modernity associated with the artificial cultural concept of masculinity is the biceps curl, an example of a third class lever. Force = biceps contraction, Resistance = Weight placed in hand, and Axis = Elbow.

CONTROLLING LEVERS

While we can classify individual muscles acting on a single joint as a specific lever type, things are rarely so cut and dried. The human body revels in compound movements, meaning that multiple joints and muscles are simultaneously in action. When considering functional anatomy, it is important to carefully consider contributing muscles and joints when doing an analysis of a movement. For example, many people think of squats as a quadriceps exercise, thus, they consider only the action of the knee joint. It is much more relevant to also consider the hip joint, as this joint contributes greatly to the movement. The musculature involved in the squat is not just the quadriceps - the hamstrings and gluteus muscles are also greatly involved. This means a simple movement, squatting, is composed of multiple lever systems, multiple muscles, and multiple joints. Simple observation can tell us this and later we will find out how to easily approach such an analysis.

During exercises we want to use the levers of the body to produce movement of body segments or the entire body while maintaining control of the body so that it does not perform extra work. Essentially we do not want to create any unnecessary lever arms that introduce movement inefficiency (bad technique).

How do we create unnecessary lever arms and how do we know if we have? To answer this let's first consider the concepts of "Center of Balance" and "Center of Gravity".

CENTER OF BALANCE

Center of balance is the point where the body contacts the ground and above which the body's mass is balanced. The center of balance is quite plastic and constantly changing as we unconsciously exert muscular effort to keep our balance and not fall over – even when we are standing still. If we are standing on both feet the center of balance will occur somewhere on the soles of the feet (both of them), usually between the calcaneous (heel bone) and the metatarsals (ball of the foot).

CENTER OF GRAVITY

Center of gravity (sometimes referred to as center of mass) is the point within the body where there is equal mass above, below, in front, and in back (the intersection of the cardinal planes). As with the center of balance, the center of gravity can change depending on the conformation of the body.

Normal standing posture usually places the center of gravity a few inches above and behind the navel. If we flex the hip and hold the foot above waist level to the front (figure 3-4) the center of gravity will move forward and be slightly anterior to the body but still roughly at the level of the navel. If we add an additional element, an external load for example, this will change the effective center of gravity of the body to include the mass of the load. A load supported above the head moves the center of gravity upwards. A load held low moves it downwards. The larger the load, the larger the deviation of the center of mass from its unloaded location.

Center of gravity of changed posture

Normal standing center of gravity

Figure 3-4. Placing the body in different positions alters the location of the center of mass/gravity.

STANDING EXERCISES

When we do standing exercises we need to control the center of mass and keep it within the boundaries of support identified by our footprints. The most efficient technique is one that keeps the center of gravity directly over mid-foot. If you can move loads in as close to a straight line as possible and as close to the line between the center of balance and the point of load application you will be moving efficiently. You will not have induced any extraneous lever arms that require additional work – movement that does not contribute directly to the completion of the movement.

What happens when you stand at attention then lean back? As you lean back your center of gravity moves back and thus moves the center of balance back towards the heel. If no balance corrections are made and the center of gravity moves behind the most posterior aspect of your support structure - you fall backwards. The opposite, a fall forward, occurs if the center of gravity is allowed to move forward beyond the most anterior support structures - the toes (Figure 3-5).

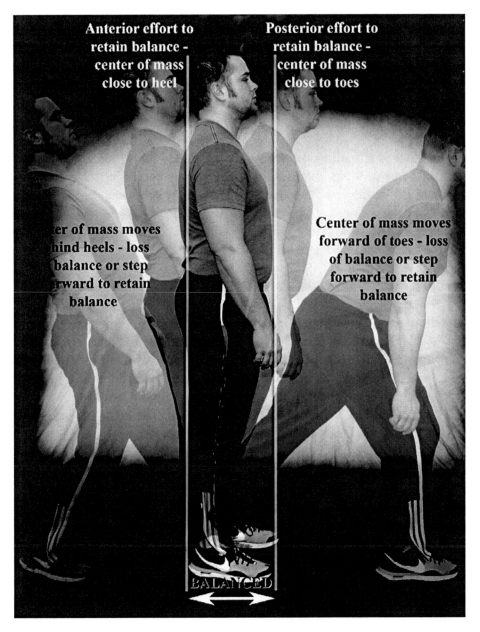

Figure 3-5. Deviations of the center of mass/gravity within the bodies active support structures are corrected with reflex muscular activity.

Luckily falling over from such occurrences does not happen with any frequency in humans as we have wonderful little reflexes that prevent it. When the center of gravity cannot fall within an anatomically defined center of balance, large-scale corrective movements must occur (step forward or step back for example).

If the center of gravity moves forward, posterior muscles like the gastrocnemius, gluteals, and erector spinae contract to pull the body backward into balance.

If our center of balance is sensed to be too far back, a nuanced series of muscle contractions begins that bring the skeletal system back in line. Anterior muscles such as the tibialis anterior, rectus abdominis, and iliopsoas contract to pull the body and center of gravity forward into balance, bringing the knee back over the ankle, the hip over the knee, the shoulders over the hips, and it all occurs in a fairly imperceptible manner.

In more unbalanced situations the arms may get pulled forward and the head may get rolled forward over the toes for a larger scale correction. What has occurred here is that the lean backward (or forward or sideways) has induced a lever arm that has produced a moment against which we have to produce force to balance the system. The fewer and the smaller magnitude corrections we have to make, the better, and the more efficient we are.

BALANCE DURING DYNAMIC MOVEMENTS

When we do dynamic movements, such as simple walking, things are a bit more complex. Walking has been described as a controlled loss of balance forward. When we want to walk forward we recruit a set of muscles that propel our center of gravity forward beyond the anterior support capability of the trailing foot. But instead of falling over, the heel of the lead foot picks up the load. More muscular activity moves the center of gravity forward. The center of balance moves from heel to toe-then beyond and the cycle repeats.

WHAT IT COMES DOWN TO

Anatomically, we are dealt a fairly static hand when it comes to levers – we are built the way we are. We can't train to change the location where muscles attach to bone in order to improve mechanical advantage. BUT we can teach efficient technique by routinely and carefully observing trainees for lever arms that have been erroneously introduced into an exercise. Once identified, we then practice the art of coaching to get rid of them. To do so accomplishes two purposes: (1) better performance leading to faster fitness gains and (2) safer exercise sessions. Accomplish these two things and we all win.

4 - FORCE THE ISSUE

"I find your lack of faith disturbing" and then squooosh, force choke! Silent, brutal, indefensible. In one dramatic squeeze of a fist, Darth Vader redefined the popular notion of what the "force" was. Albeit a really cool mystical entity, the "force" in application still falls under the same basic laws of forces and physics that the rest of us mere mortals are resigned to live under.

FORCE DEFINED

Force is a phenomenon that causes an object to move, accelerate, or that produces stress in an immovable body. A force, by definition has a magnitude (how big the force is) and a direction (which way it is pushing or pulling), making it a vector quantity. Vader's force choke indeed had a magnitude, large enough to crush a human trachea, and it had a direction, let's assume that the force choke acted like invisible fingers being squeezed toward the heel of the hand. It's fairly amazing that Isaac Newton had described the physical realities of mythical forces in the Star Wars universe by 1687. Force generation and application is how we move our bodies. The rules surrounding forces and their application have been known for centuries.

In modern physics we can precisely quantify forces in a variety of units (newtons, dynes, kiloponds, pound-force, or foot-pounds) and we can determine in which direction the force is acting. This quantifiability makes force measurement an attractive means of assessing efficient human movement, fitness level, and even forms the basis of many sports.

FORCES RELEVANT TO EXERCISE

The type of force and it's method of application to the body are extremely germane to understanding how the body moves during exercise, information that is useful in knowing how to coach movement. Different types of forces act on human anatomy in different and specific manners. Conversely the human body is constructed to accommodate these forces and to adapt to changes in the nature of their application to ensure survival. Forces acting on the body, thus test the limits of existing anatomy while simultaneously driving changes in anatomy. We can divide the basic types of exercise-related forces as such; compression, tension, torsion, and shear.

Compression

If you put a barbell on someone's shoulders, gravity is going to pull the weight toward the ground with a force that is applied along the long axis of the body. In other words the body is being squished between the weight and the earth. It's a good thing for us that the body is built precisely to tolerate huge compressive forces. After all, supportive bones such as the vertebrae have similar compression resistance to oak wood structural beams. So it is relevant to note here that in healthy trainees, compression of the skeleton rarely leads to structural failure (things breaking). Paul Anderson stood up with over 6000 pounds on his shoulders. Bill Clark at age 70 did a hip lift with in excess of 1000 pounds, both huge compressive forces that produced no anatomical damage.

Proper exercise technique in the face of large compressive forces prevents injury because it allows bones to support the weight, not the less rigid muscles, tendons, and ligaments around joints. Obviously these latter structures contribute, but proper alignment of skeletal elements during exercise allows distribution of compressive forces to those structures that are most resistant and thus prevents orthopedic problems (and bone on bone movement provides more effective force transfer as a perk). Exercises that possess compressive characteristics, when done progressively, develop additional tolerance to compressive force. Exercises with a compressive nature, when done for months to years stimulate bones to add new mineral content thus reinforcing the bones making them even more resistant to fracture.

Tension

The opposite of compression is tension. If someone hangs on a high bar, gravity is going to act on the body by attempting to pull it down to the ground. The grip on the high bar prevents this from occurring, but the downward force along the long axis of the body remains. This is tension. You can think of tension as a stretching force. Adding weight to the system (weights on a hip belt & chain for example) is one way to increase tension, but adding movement to the system can also increase tension. In the performance of a back hand-swing on a gymnastic high bar, a child can experience a stretching force along the long axis of the body of nearly six times gravity (6G). Again, the human body is built to handle application of such linear forces. Our ligaments represent a safeguard, whose engagement represents the limit of space between the bones comprising a joint (exceeding that limit represents ligamentous injury). Vertebral ligaments, once engaged, can stretch about 20mm before disruption and a single intervertebral joint will withstand about 224 pounds of tension before failure. Note that this value is from a cadaver study thus under-estimates the force required to disrupt a living joint by about 25%. The value also represents an un-resisted load whereas in the living body the vertebral musculature would produce a counterforce making the actual force required to produce an injury significantly larger. Regular progressive performance of exercises that produce tension drives gains in tolerance as the loaded ligaments become stronger and thicker but more importantly the musculature that counters the effects of tension become stronger, enduring, and more effective in maintaining joint integrity.

Figure 4-1. Compression occurs when the body is between an immobile surface and a load attempting to move towards the immobile surface as shown here OR when the body is in between to loads moving in opposite directions.

Figure 4-2. Tension occurs when gravity acts to pull a suspended body towards the ground as shown here OR when two loads moving or attempting to move in opposite directions and place a stretch on the body or a body segment.

Torsion

If we put a wrench on a nut and then pull on the wrench, the force is applied to the nut in a rotational manner. We can call it torsion or torque (ever heard of a torque wrench?). The same thing occurs in most instances during human movement. We can induce torsion on the vertebral column simply by selectively contracting our own muscles to move our shoulders left or right. Torsion is a normal part of human activities and inclusion of torsional exercises or exercises that develop the ability to produce or resist torsion is an appropriate training strategy. We need to prepare our bodies beyond producing and resisting linear movement. In many instances we can be recipients of involuntary torsion. A gut wrench in wrestling places a large amount of torsional stress on an opponent with the express intent of forcing them to rotate their body to expose their shoulders to the mat. Torsion applied to joints not designed to move in a rotational manner is a rather common problem in plant and twist sports such as basketball, soccer (football), and football (American) where we see a large number of ligamentous knee injuries from rotating the femur on top of an immobile tibia. It may not seem intuitive, but comprehensive strengthening of the knee musculature has been shown in many studies to aid in reducing the frequency of this type of injury, most likely due to a stronger set of muscles keeping the joint within its normal ligamental limits. For precision, we need to acknowledge that the preceding is a conceptual presentation. The nut described above rotates on an axis and although the force applied to the nut through the wrench produces rotation, the more accurate term is "moment". The distance between the point of resistance (the nut) and the point of force application (the hand on the wrench) is called a moment arm. The longer a moment arm, the larger the moment at the axis of rotation. This will be useful when considering levers and movement.

Figure 4-3. A load placed on the body that applies a rotational force is torsion or torque. In the example here (A), during the "good morning" exercise the bar is pulled down by gravity and torsion occurs around the hip, the axis of rotation (or fulcrum of the system). Exceeding the ability of a joint's, or system of joint's, musculature to maintain stability results in bending as in panel B, or other possible movement or postural errors.

Shear

If one segment of the body has force applied to it in one direction and an adjacent segment has force applied to it in an opposite direction, there is a shear force present. The presence of shear is not particularly dangerous as long as the musculature involved is developed to the point that it is strong enough to prevent movement of the adjacent segments in opposite directions. Un-resisted vertebral shear could pose significant problems. Thus, exercises that strengthen the muscles that stabilize the vertebral column are vital to fitness, sport performance, and avoidance of orthopedic injury. However, even when the shear force is un-resisted, it takes about 336 pounds of pressure across one intervertebral joint to induce failure and bone translation (linear sliding) in cadaver studies. Again, in the living human it requires a greater un-resisted force to induce joint failure and when muscular contractions produce stabilizing forces, the amount of force required for failure would be even higher.

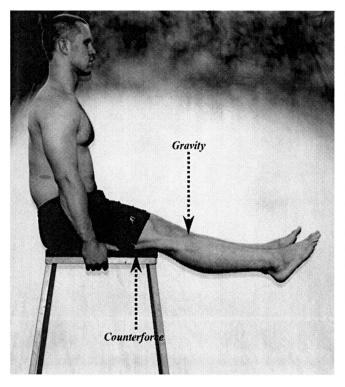

Figure 4-4. Shear occurs when two segments of the body are pushed in opposite directions. In the example here the table provides an upward force on the thigh and gravity provides a downward force on the unsupported lower leg. This places a shear force on the knee. If the complete musculature surrounding the knee is well developed this is not a problem, if an injury exists or in conditions of extremely low fitness shear force can produce shear (movement) and this may be a problem.

ROLE OF MUSCULAR CONTRACTIONS.

Any of the forces described can be produced by muscular contraction alone, in concert with or against gravity, or acting together with other bodies. And that is an important concept to understand. The body can experience these forces from external application AND it can produce the forces described through contraction.

Forces across a joint - Let's turn to another issue related to force production, the balancing of forces across a joint. In most instances during exercise we want to create, through muscular contraction, an unbalanced force across a joint or joints in order to create movement. If I am down in a squatting position and I want to stand up I have to generate a force with my hip and knee extensors that is greater than the effects of gravity on my body mass. If the force is not unbalanced in the direction opposing gravity, either because of weakness or because of an additional load on my shoulders, I cannot stand up. That is usually a bad thing (not being able to stand up).

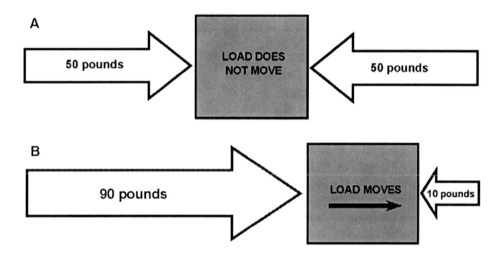

Figure 4-5. In the balanced force case here (A), equivalent and opposing forces cannot induce movement. Only with an unbalanced force (B) scenario where one directional force has a magnitude greater than the opposing directional force can a load be moved. The greater the difference in magnitude between the opposing forces, the more impulse available to move the load quickly.

When we consider joint integrity during movement, it is often proposed that we need to balance forces across the joint in order to stabilize it. This point is interesting; we do need to counter any force that actively attempts to engage ligaments at their limits. Remember that muscles and tendons try to keep joints pulled together, ligaments act only to prevent them from separating. For example, having the hamstrings and quadriceps co-contracting effectively during squatting prevents forward or backward movement of the knee joint (shear between the femur and tibia). This is a good thing (preventing shear). But how do we manage to produce an unbalanced force that is driving movement and then balance this force across a joint? That brings us to the final order of business, a consideration of agonist (primary mover) and antagonist (resists the action of the primary mover) muscles and the mythical optimal strength ratios.

Mythical optimal strength ratios - It is common to see suggestions that there is clinical and exercise training relevance produced by dividing things like hamstring strength by quadriceps strength to obtain an indication of muscular balance. It is frequently suggested that the agonist-antagonist ratio should approach 1:1 in rehabilitation and exercise settings. The International Fitness Professionals Association has provided strength (force generating) standards for its members (table 4-1).

Muscle Groups	Balance Ratio
Ankle Inverters & Everters	1:1
Ankle Plantar Flexors & Dorsiflexors	3:1
Elbow Flexors & Extensors	1:1
Hip Flexors & Extensors	1:1
Knee Flexors & Extensors	2:3
Shoulder Internal & External Rotators	3:2
Shoulder Flexors & Extensors	2:3
Trunk Flexors & Extensors	1:1

Table 4-1. International Fitness Professionals Association's Strength Balance Standards.

A problem with this concept is that it is not reflective of real-world movement. The leg extension and leg curl exercises, involving only the knee, are the usual tests of lower body muscular balance (since every fitness club and clinic has these machines). These two movements do not occur during normal human ambulation, sport, or during exercise movement without being accompanied by other joint movements. The same can be said of the other muscle groups listed and their methods of assessment. And more importantly, a second problem is that these ratios are frequently interpreted to imply that the opposing groups are active during activity at the ratio listed. This leads to a misunderstanding of the fact that any movement of the human body, by definition, must be the result of an unbalanced force and the magnitude of force produced by the agonist is irrelevant to a predetermined ratio.

Another problem with ratios is that they are generally based on opinion and defensive medical thought. Very little quality research has been done in this area. Weight for the lack of usefulness of ratios can be found in a brief statement regarding the relevant literature by Dr. Jonathon Fass (PT, DPT, ART, CSCS) published in *ADVANCE for Physical Therapy and Rehab Medicine*; "the scientific literature has had limited success identifying the link between strains/tears and weaknesses in synergists or strength balances between agonists".

Agonist/antagonist force production - A balanced force across a joint is not the same as "strength balance" between agonists and antagonists. A precisely balanced force across a single joint would result in an isometric muscle action and produce no movement. Antagonist muscles are not recruited extensively, nor are they inhibited in force production in order to allow the agonist to drive movement. In fact, one of the first and basic physiological adaptations the body makes in the first few months of training is that we learn how to turn off antagonist muscles more completely in order to maximize the efficiency of the agonist muscles in executing a movement. Less resistance from the antagonist means more available force from the agonist. This renders the 1-to-1 (or 2-to-3 or 3-to-1 etc) concept of strength balance irrelevant during movement. If it takes 25 kilograms of posterior force to maintain knee structure stability during movement, does it make any difference if the muscles providing that force can, at maximum effort, produce 50 kg of force or 500? Not really. There will obviously be a minimum level of force required to maintain structural integrity of joints, and stronger is always better, but the ratio of agonist/antagonist force production during movement is a function of physical task requirements and individual anatomy.

This does not mean that an antagonist muscle may not be active during agonist contraction. Think about the hamstrings (biceps femoris, semitendinosis, and semimembranosis) during standing up out of the bottom of a squat. The hamstrings are normally thought of as antagonists to the quadriceps (vastus medialis, vastus lateralis, vastus internus, and rectus femoris). They are quite active in this movement but not as antagonists to the knee extensors, they are themselves agonists in hip extension. That is their job. That is the job of all muscles, to produce, not resist large-scale joint movement. In our squat example, the traditionally held antagonist muscles simply generate the necessary joint stabilizing counter-forces, as part of their roles as prime movers. This is an involuntary by-product of basic anatomy, physiology, and physics.

Antagonists are important, do not get me wrong; they assist in deceleration, stabilization, and fine motor control, but the notion that we can magically determine how strong the multitude of agonist and antagonist muscles need to be, as well as the ones active in simple exercise is not practical. Measurable estimations of the relative contributions of individual or groups of agonists and antagonists during complex movements are not well known, especially in fit populations. So how can we construct a hierarchy of concept and terminology based upon something that is so poorly understood?

A one-size-fits-all statement about optimal strength balance is unrealistic, and if someone does per chance hear or read about one, know that it cannot provide us with any practical advantage when coaching in the gym. It is our charge to develop our trainees to be fully functional, able to both tolerate and produce a complete spectrum of real forces. We need not minimize the forces presented during training or tune force generation capacity to a hypothetical ratio.

The best advice is to strengthen all relevant axes of movement around a joint. For example, if you bench press, then you need to press, dip, chin, and do deadlifts or rows to "balance" the strength of the muscles around the shoulder joint.

If a complex movement develops force generation capacity in multiple axes around a joint then forgo isolation exercises, as strength balance will be naturally obtained during the execution of the compound exercise (think of how this happens in the squat).

Where force production is concerned, we need to focus on programming exercises to improve force production, force tolerance, and absolutely of most import, functional fitness enhancement for the real world.

"The human head is of the same approximate size and weight as a roaster chicken. I have never before had occasion to make the comparison, for never before today have I seen a head in a roasting pan."

- *Mary Roach*
in Stiff: The Curious Lives of Human Cadavers

5 - RATTLING CHAINS

A great deal of misunderstanding exists about kinetic chains. Every movement has one. Clinicians place a great deal of import on assessing them. And sport performance depends on their effective function. But what is a kinetic chain? And do we really care if we don't know what one is? A kinetic chain is simply the collection of body segments that contribute to a movement. More precisely the kinetic chain is the assemblage of muscles and bones that makes the body, or part of it, move in a certain way. If I do a bodyweight squat there is a different kinetic chain operating than when I do a handstand push-up. Different muscles and different bones are involved in each of these exercises, thus two different kinetic chains are operating.

The skeleton is an important part of the kinetic chain as it allows for transfer of force from segment to segment, but the musculature actually contributing to the motion (or force generation) is the real meat of the matter. Muscle makes movement. Therefore, frequently we only think of muscles as the kinetic chain. This is OK since we are exercising our muscles to develop them in order to improve the function of the kinetic chain, but it must be understood that bones and joints are essential parts of the chain.

OPEN VS CLOSED KINETIC CHAIN

Analysis of a movement and its kinetic chain is usually done in too simplistic and arbitrary of a manner. Frequently the only level of analysis is to classify the movement as "open kinetic chain" or "closed kinetic chain". These terms are very popular in the fitness and clinical exercise professions where isolation exercises are frequently the vogue.

Open kinetic chain movement – The distal end of the moving body segment is mobile, as in the leg extension.

Closed kinetic chain movement – The distal end of the moving body segment is immobile, as in the squat.

At this level of analysis it is pretty easy to say that open chain exercises are essentially isolation and machine exercises that have limited transfer to real-world improvement of function and that closed chain exercises are usually compound or free-weight exercises that provide the best transfer to functional enhancement.

But it's tricky. If I do a press (a standing, over-head, barbell press) the distal end of the body is moving – so its an open chain exercise. Right? Well, let's think about that. In the squat, the feet are considered the distal end and they are fixed. The bar is on the shoulders and is moved by the motion of the legs. Muscles in between the feet and the bar move the bar. How is this different from the press where the feet are fixed and the muscles of the shoulders and arms move the bar? Its not really too much different at all. The body, which is wedged in between the earth and the weight, is causing the weight to move. The only difference in the kinetic chain is its location - arms vs. legs. Sometimes in the exercise and clinical sciences, things don't get thought out too well. Sometimes in the quest for precision, things become less clear. Here it is in a nutshell:

Open Kinetic Chain Exercises

If you're sitting down and doing your isolation exercises with your hands and feet waving around in the air while being connected to the world through your gluteus maximus and ischial tuberosity interacting with a comfy padded seat on an exercise machine or exercise bench, you are doing open kinetic chain exercises.

Closed Kinetic Chain Exercises

If you are standing on your feet supporting a free-weight through the length of your body, you are doing closed chain exercise. There are a few exceptions but in general this is the crux of open and closed chain exercise. Not too important a concept, since we already know that isolation exercises are less effective that large scale free-weight exercises in improving strength and overall fitness.

THE UNIQUE KINETIC CHAIN

Instead of assigning movements to arbitrary categories, let's think of each movement as having a unique kinetic chain, not open, not closed, just a kinetic chain (actually the only legitimate use of the terms "open" and "closed" in describing movement is in reference to joint angles - as is the case in gymnastics).

Magnitude and Location

It is probably most useful to think of the magnitude of the chain. A larger chain involves more muscles, expends more calories, implies a larger training stimulus, and thus delivers better fitness gains than a small chain exercise. It is also useful to think of where the chain is. Posterior kinetic chain exercises centering at the hip greatly contribute power performance more so than anterior kinetic chain exercises centering at the knee because of magnitude of muscular involvement.

Combinations

Exercises combining both are even more effective – more joints, more muscles, bigger chain. Let's think about the press again. How large is the kinetic chain? If you think about it, every muscle and joint between the soles of the feet and the palmar surface of the hands contributes to movement, either through direct movement of the bar or through stabilizing the chain (entire body) to facilitate efficient force transfer. So when I press I am actually developing my rectus abdominus, gemellus, gastrocnemius, flexor pollicus brevis and more.

It must be understood that considering only the moving segment is frequently not useful in analyzing a movement since there are so many other contributory anatomical elements. Yes, the primary movers, the deltoideus muscles, the triceps brachii, and others are getting the brunt of the work, but the rest of the stabilizing and contributory musculature is also working and getting more fit.

Holistic approach

Get in the habit of thinking holistically about movement, about muscles, and about joints. Don't get caught up in jargon and an isolationist approach to exercise analysis. You are much better off as a practitioner knowing the muscles, the bones they move, and how they all work together to move the body singly and in concert than you are knowing the specious little terms constructed to make the study of anatomy and human movement sound more important and difficult than it already is.

Figure 5-1. Examples of kinetic chains. The magnitude of difference between the length of the kinetic chain and muscles involved in the squat (left) and the leg press (right) is quite large owing to the distance between the exercise support surface (the floor vs. the back rest) and the application of resistance (where the weight is). It should be apparent that these two exercises are clearly different in the nature of training adaptation they will produce based on this simple observation.

"No man should marry until he has studied anatomy …"

- Honore de Balzac

6 – BONING UP

Bones are the anatomical elements most associated with the study of anatomy. Every medical clinic, every high school biology department, every university program where the human is studied, and heck, even most spas, fitness clubs, and gyms have at least one skeletal chart or model somewhere (frequently though the last three have them more for decorative purposes than anything else). Exercise professionals need to know what bones do and how they are built individually and as part of a larger system.

Bones do a lot and there are a lot of them – 206 to be fairly precise (there is a small variation in number among ethnicities and individuals), almost all of which can be seen in figure 1-6. Bones are hard but alive; they grow and adapt. And very much at the heart of the matter here, bones provide the hardware for human movement. They act as internal scaffolding supporting everything visceral and muscular so we don't look like a pile of quivering protoplasm. They form the lever arms that move us and allow us to move other things. They act as implements to strike things and as protection from things that strike us. They also provide us with a very important component of the cardiovascular system – blood cells.

The study of bone is frequently called "osteology" but we're just going to say we are studying the anatomy of bones. Extra ten-dollar words that don't mean much to the average Joe can't help us coach anyone better.

Even though we can simplify the process of learning anatomy by eliminating unnecessary vocabulary, it is never truly an easy task. To this point we have been discussing concepts of anatomy. We are embarking on the part of anatomy everyone hates, memorizing structures. Lots of people say that they have an easy way to memorize things. You need to realize that your brain works fairly uniquely and that those techniques may not work for you. One method of learning and memorizing that does work, always, is repetition. When you learn finger position for a C chord on a guitar, how did you master it? Repetition. When you learned your cell phone number, how did you memorize it? Repetition. When you learned to tie your shoes how did you accomplish that? Repetition. These three examples represent a purely physical task, a purely intellectual task, and a combination task. Each requiring some form of memorization. Anatomy is no different, if you can do the above, you can learn anatomy. The only keys to success are time and repetition.

BASIC BONE STRUCTURE

The first task is to examine the basic structure of bones. There are several distinct types: (1) Long bones, (2) Flat bones, (3) Short bones, and (4) Irregular bones. Of the four types of bones, long bones are likely the most important to sport and exercise as they provide for the lever systems that make movements possible. Named "long bones" simply because they are the longest bones in the body, they have a characteristic anatomical structure (figure 6-1). Areas of anatomical importance in long bones are the epiphysis, the epiphyseal plate (commonly known as the growth plate), the diaphysis, and cancellous bone.

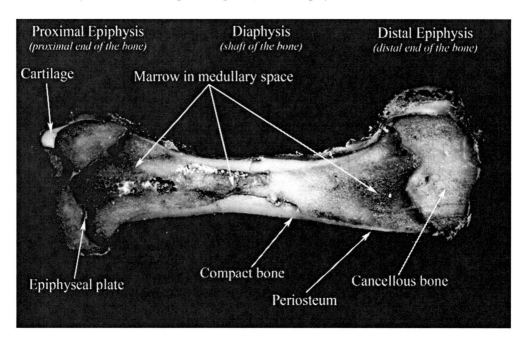

Figure 6-1. A trip to the pet aisle of your grocery store will yield a nice example of a long bone. For a couple bucks or so you can pick up a swine femur and explore bone structure. If you happen to have a radial arm saw handy you can cut the bone longitudinally to reveal its inner structure.

ARCHITECTURAL ELEMENTS OF BONE DEFINED

Cancellous Bone - This feature of long bones has also been called spongy bone as it appears similar to the texture of a very tight sponge. In reality, there is a lot of material packed into the sponge and it may be more like hard wet sand with many many visible pinholes. Cancellous bone is usually found closer to the epiphysis and functions to help absorb external mechanical stress.

Cartilage - This is a fibrous tissue that lines joint capsules where bone meets bone. The more a bone end is exposed to physical stress, the thicker the cartilage will be.

Compact Bone - This is layered hard bone found in the diaphysis (shaft). Compact bone is comprised of mineralized connective tissue and is a resilient and tough structure that gives the skeleton much of its ability to support and protect.

Diaphysis - The term diaphysis is used in reference to the shaft of the bone. It is made of hard, compact bone and has a central medullary cavity or canal.

Epiphysis - The term epiphysis describes the end of a long bone (there will be two of them). They are primarily composed of cancellous bone.

Epiphyseal Plate - The epiphyseal plate is a thin but diffuse layer of cartilaginous tissue located between the epiphysis and diaphysis. There are many mitotically active cells here and it is the site of bone growth. At full maturity, the cells cease activity and the plate fuses with the existing bone. At that point longitudinal growth (gain in height) stops.

Medullary Cavity - This central cavity is the site of bone marrow genesis and maturation in the young. Later in life, fat largely replaces much of the marrow. Marrow is responsible for blood cell production. In figure 6-1 the tissue identified as marrow in the medullary space appears as a gritty gel like substance to the touch. You can use your finger to remove the marrow and expose the cavity. The marrow is what your dog is trying to get at when it gnaws on a bone.

Periosteum - The periosteum appears as a fibrous sheath surrounding the bone. It has multiple functions, one of which is to provide nourishment to the bone. It is also the site of developing bone cells (osteocytes).

SMALL FEATURES OF BONE

There are individual names for each bone and learning them is easy enough. However, there are little bumps, ridges, depressions, holes, knobs, and more on a bone that also have specific names and specific functions. We'll call these bone parts "small features" as they are small compared to the bone upon which they occur. This adds to the complexity of the learning task. A few basic conventions, mastered now, will help later as the rules for naming these small features are applicable throughout the skeleton.

Tuberosity, Tubercle, or Process - These three names describe similar bony features that appear as an obvious lump, bump, or elevated area on a bone. The difference in their sizes is noted in figure 6-2. These bony protuberances serve as sites for tendonous attachments to the bone and are the points of muscular force application within the body's numerous lever systems. Frequently there will be multiple muscles that attach to a tuberosity and that can selectively be recruited to move the bone upon which the bump is located in various directions.

Notch or Groove - Most often this feature appears as a short "notched out" area of bone that generally serves as a pathway for nerves, vessels, or connective tissue. This gets the sensitive tissue out of the way and protects it within the recesses into the rigid structure of the bone (figure 6-3).

Spine - This is not the term commonly used to identify the "back bone". Rather it refers to a spine, an obvious projection from a bone (figure 6-4). It can be discriminated from a tuberosity in that it is generally longer and thinner than a tuberosity. The little bumps you can feel along your backbone are spines of individual vertebral bones.

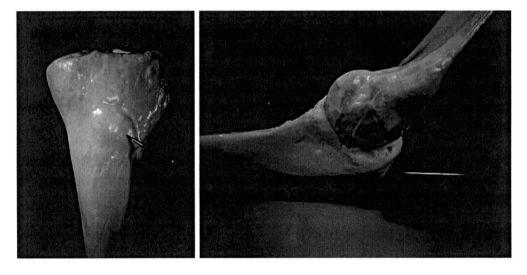

Figure 6-2. The tibial tuberosity and the olecranon process compared. It is worthy of note that a tuberosity (tibial tuberosity - left) is larger than a tubercle, which is larger than a process (olecranon process of the elbow - right).

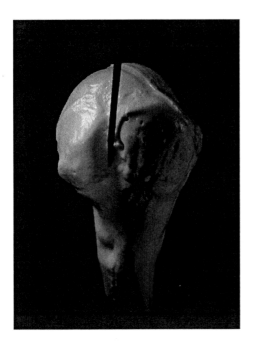

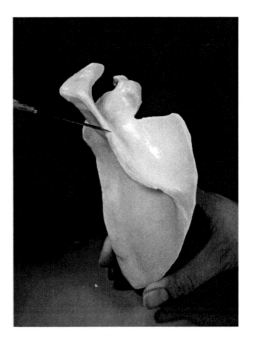

Figure 6-3. The bicepital groove of the proximal end of the humerus.

Figure 6-4. The spine of the scapula. This is an extremely important bony feature as all activities carried out with the arms funnel their forces through this feature.

Fossa - A fossa is a larger concave shaped area of a bone (figure 6-5). It may or may not be obvious upon first inspection. Some fossa are relatively shallow bowl shapes. In some instances the term may be used in reference to concave surfaces on the body (apparent on the skin but resulting from the shape of the underlying muscle, tendon and bone). The antecubital fossa on the inside of the elbow (where blood is drawn from) and the popliteal fossa, the depression on the back of the knee, are examples of this latter concept.

Foramen - A foramen is simply a hole passing completely through a bone (figure 6-6). Nerves, muscles, and blood vessels may pass through a foramen. Foramen vary is size from small to very large.

 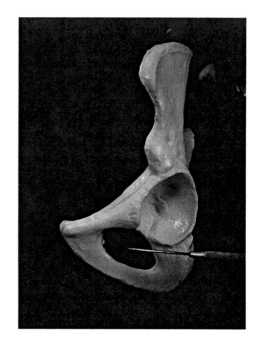

Figure 6-5. The glenoid fossa of the scapula.

Figure 6-6. The obturator magnum of the pelvis. Not all foramen are this big.

Condyles - These are smooth knob-shaped features that form the articular surfaces (where joints come together) of the bones of moving joints. The condyles of the distal femur are shown in figure 6-7. You may hear condyles referred to the knuckle of a bone.

Crest - This is a superior, narrow, ridge-like portion of a vertically aligned bone, as seen in figure 6-8.

Linea - This is a long raised area along the length of a bone serving as a site of muscle attachment (figure 6-9). The term can also be used as a name for a line of tendon embedded in a muscle (linea alba of the rectus abdominis).

Trochanter - This is a large raised area of a bone where a large tendon (tendons) and ligament(s) are attached (figure 6-10). A trochanter is much larger than a tuberosity, tubercle, or process.

Epicondyle - An epicondyle is a smooth outcropping of bone just superior (and lateral or medial) to a condyle. The medial epicondyle of the humerus is illustrated in figure 6-11.

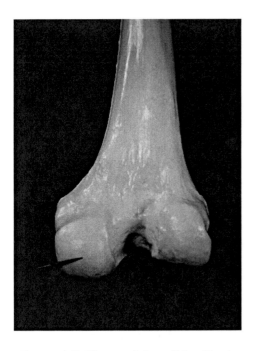

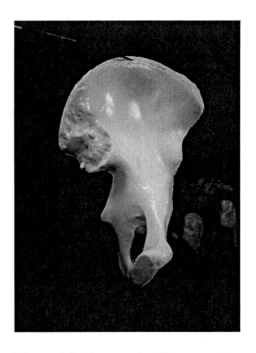

Figure 6-7. The condyles of the distal femur.

Figure 6-8. The superior iliac crest of the pelvis.

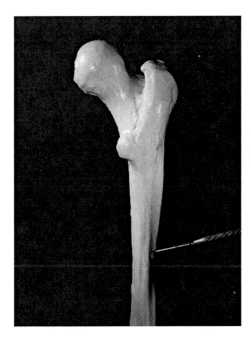

Figure 6-9. The linea aspera of the femur appears as an elevated surface running vertically on the bone.

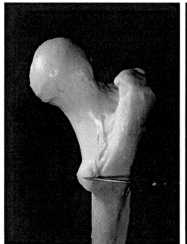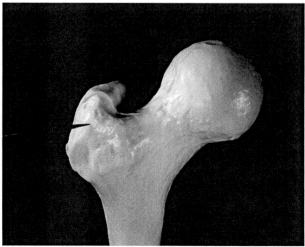

Figure 6-10. Trochanters are larger outcroppings of bone. Sometimes there are more than one of the same type feature on a bone. In that case, a modifier will be added to the name, for example the lesser trochanter (left) and greater trochanter (right) of the femur.

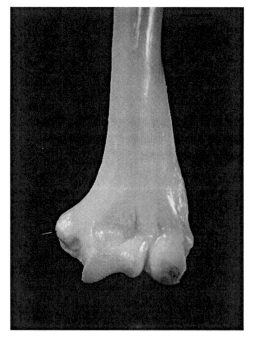

Figure 6-11. The medial epicondyle of the humerus. There is also a lateral epicondyle in the photograph (hint it's smaller).

THE VALUE OF KNOWING

Localizing the Site of Pain

In the words of Austin Powers, "So what does it all mean, Basil?" Of what use is knowing these small features? First, these bony features are generally palpable, meaning that we can feel them through the skin. This simple ability means that we can figure out which muscles are where and what they are doing relative to the bones and bony features to which they are attached. In the case of trainee muscle pain, you can use the information to localize, to a specific muscle or muscles, the site of pain. You would then be able to possibly identify the precipitating injurious event AND more importantly select exercises for use during the initial healing process that reduce or temporarily eliminate the load on the injured muscle(s).

Developing a Safe and Effective Exercise Technique

And then there is a strong case to be made for knowing what muscles do. If you do not know that a muscle crosses two joints and has both a proximal and distal function, what are the chances that you will be able to develop an exercise strategy that fully develops that muscle's complete function? What are the chances that you will be able to teach correct, efficient, and safe exercise technique without such understanding? Yes, you may be able to mimic someone else's teaching method and get it right, or maybe you just might get it right by chance. But if you plan to be a professional and you plan to be able to recreate training success, it is a wise move to know why you do what you do. Not only for the purpose of maximizing trainee fitness gain but to also be able to explain the why's of exercise to that same trainee.

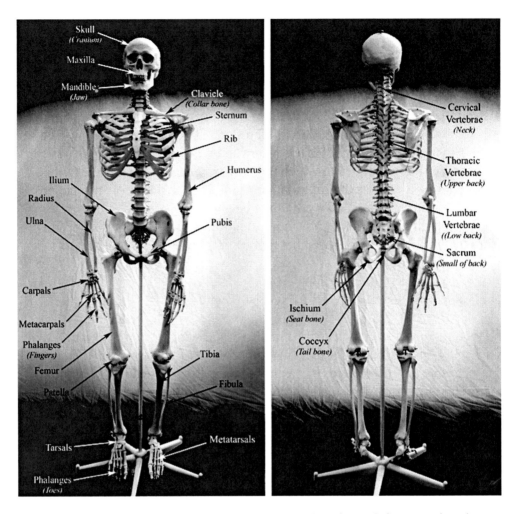

Figure 6-12. Basic bones of the human body. Anterior view – left. Posterior view – right.

7 - GYM JOINTS

Every move we make and every load we support is accomplished with the aid of at least one joint. Joints (articulations) take many forms. From virtually immovable connections between the flat bones of the skull to the wildly moveable joint of the scapula and humerus (the shoulder joint) and everything in between. Joint anatomy is important to exercise professionals, as understanding their structure aids in understanding the finer technical points of teaching and performing exercise skills. It is important, right off the bat, to know that one MUST consider the structure and functions of all joints used during an exercise skill when teaching, learning, and executing the skill not just the one that seems to be the most dominant. Further, we need to include the geometric associations between the active joints relative to force generation in our assessments.

UNDERSTANDING JOINT ACTIONS

The analysis of muscle functions at the joints may appear confusing and difficult at first. Nothing about movement is black and white at all times. As the tasks you ask the body to do change, the nature of how muscles act in concert with one another and the innate lever systems also change. However, there are a couple simple concepts can make the process easier to approach. Understanding what a "stabile structure" and a "line of action" is will arm you with the ability to determine what will happen when a muscle contracts and flexes a joint.

Stabile Structure

A stabile structure is the non-moving end of a joint system, the end of the system around which or to which other structures are moving. If a tie a rope around a tree and then pull as hard as I can with my arms and shoulders, which will move, my body toward the tree or the tree towards my body? The tree is firmly entrenched into its position and is more stable than my body, as such my body will be pulled toward the tree. The same think occurs with the skeleton and muscles. Muscles are arranged and recruited for tasks so that one muscle is more stable than the opposing one and a correct movement is the result of contraction.

The study of muscles is traditionally approached by learning their origins (beginning) and insertions (end) and then by memorizing the functions attributed to them. In general, as a muscle contracts (shortens), it pulls its insertion on another bone toward its origin. This type of analysis and learning give us a couple functional problems, one indirect one in application and another in understanding of function. The former application problem stems from using a method where we study one muscle, one origin, one insertion, and one function. This leads to an isolationist perspective to muscular function and provides an artificial construct on which training is based – the evolution of isolation training. The problem regarding function also relates to the concept of single muscle-single function convention. If no external load (resistance) is placed on the body and the movements are carried out in very specific situations, one muscle-one function may represent reality. However, if an external load is placed on the muscle, as frequently is the case in sports, the picture changes. An external load that is greater than the force the muscle is capable of producing will cause the origin of the muscle to move towards the insertion. This renders the concept of origin and insertion less than useful. So, in sport it is germane to think of motion in respect to the most stable structure, the one that doesn't move; it makes little difference as to which site is the origin and which is the insertion. Put very simply, a muscle can only contract and when it does, it will be the least stable end of the system that moves towards the most stable one – regardless of anatomical convention.

Think of the function of the biceps brachii in the barbell curl. It functions to move the weight from below waist level to shoulder height by rotating the bar in the hands around the elbow. In this example, the shoulder, upper arm, and elbow are the most stable; the forearm and hand are the least stable. The end result of contraction is a barbell curl.

Counterpoint this to a planch in the gymnastic floor exercise discipline. The hands are immobile on the tumbling floor, thus are stable, and the shoulder and upper arm move in response to biceps brachii contraction, as they are less stable.

Generally, the most stable structure has the larger mass of the two structures and is therefore easy to identify. In a pull-up my body is less stable than the pull-up bar attached to the earth so my body moves up in response to muscular contraction. In a lat pull-down the weight stack is less stable than my body that is wedged into the machine bolted to the earth. So, by thinking about which structure is the most stable you can determine which anatomical site will move as a result of muscle contraction. We still have to know attachment sites but simple analysis of movement does not require such constraints as defining muscle origins and insertions.

Line of Force

The line of force is the direction that a muscle or group of muscles is pulling in order to alter joint position. It is dictated by the direction of muscle fiber and tendinous attachments to the bones involved. It is the line along which a muscle will generate force and induce movement in the least stable structure. This is useful information that will help to understand exercise and sport skill technique.

To determine this line, localize the muscle and tendons involved and visualize a line across the joint from the least stable muscle attachment to the most stable muscle attachment. Once this is done you can easily identify and name, with anatomical terms, the expected motion resulting from muscular contraction. In the earlier examples using the biceps brachii, the line of force was directly over the anterior portion of the upper forearm (elbow) running up to the shoulder so contraction must therefore produce flexion of the forearm. See how easy it is? You can do this level of analysis for every muscle and joint in the human body. But this is an incomplete analysis. It is important here to realize that human movement rarely, if ever, occurs with such isolation – involving only one muscle and one joint. Consider all the musculature involved in a movement not just a single muscle, and all the joints for that matter. Most people think of the curl as "the" bicep exercise and that it is the only muscle targeted when there are others that contribute to execution of a curl. Flexion of the elbow joint is not solely a bicep exercise there are other muscles involved AND the bicep crosses two joints, the elbow and the shoulder. Flexing only the elbow joint leaves its proximal function, raising the humerus to the anterior, completely out, limiting the degree of development possible.

A proper anatomical analysis includes all the muscles and all the joints involved in a movement not just a muscle or two that are commonly thought to drive the movement. It should be apparent to all that the squat is not just a "quads" exercise. Properly done it recruits the hamstrings, glutes, and adductors as prime movers along with a number of other contributory muscles. Remember that exercise and sport technique are dependent upon correct anatomical analysis when we teach or correct form. Every coach, personal trainer, and therapist must understand the big picture of movement and not adopt an isolationist perspective of one muscle-one joint-one function.

JOINT STRUCTURE

We all basically understand what a joint is, its where two bones come together. And in fact a joint is just that, an articulation between two or more adjacent bones. It may either provide stability or mobility, depending on the combined architecture of all of the joint elements. A stable joint allows little or no movement. An unstable joint allows a wider range of motion depending on the overall joint structure. There are a number of contributors to joint structure and stability; ligaments, soft tissues such as muscle and tendon, and joint architecture (arrangement of the bones comprising the joint).

There about 40 different names for differing joint types there is even a name for the study of joints, arthrology. We will confine our studies for simplicity as we can get a very representative concept of joint structure and function of all joints by examining three basic classes of joints. The three basic types of joints are: fibrous joints, cartilaginous joints, and synovial joints.

Fibrous Joints

Fibrous joints are very stable and allow no ***observable*** movement – the flat bones of the skull (which actually have a degree of non-flatness) are joined by seventeen suture joints (figure 7-1). The amount of contact between bones in this type of joint provides a very stable and protective environment for the brain while allowing for a degree of resiliency of the skull upon impact. In the skull the suture joints are comprised of the bones joined and Sharpey's fibers that connect them. The derivation of this particular set of joint's name, sutures, is based in surgical terminology, as the appearance of the joints resemble the configuration of surgical sutures (the stitching that closes incisions). There are fibrous joints between the tibia and fibula, radius and ulna, and between the teeth and the jaw.

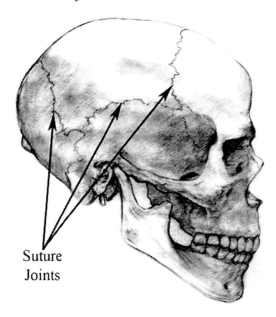

Figure 7-1. Suture joints of the skull.

Suture Joints

Cartilaginous Joints

Stability is also a characteristic of cartilaginous joints. However, unlike suture joints, limited movement is possible in cartilaginous joints. This type of joint is comprised of the bones to be articulated and the connecting cartilage (fibrocartilage or hyaline cartilage). Intervertebral disks (Figure 7-2) are excellent examples of this type of joint. They provide stability and protection for the spinal cord within the vertebral column while still allowing for spinal flexion, extension, and rotation. The pubic symphasis is capable of very limited motion compared to the intervertebral joints but belongs to the same class of joint.

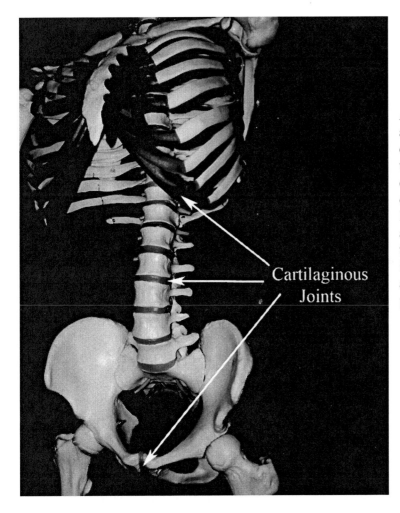

Figure 7-2. There are a number of cartilaginous joints in the skeleton. Three can be seen in this one photograph. From top to bottom: articulation between the sternum and ribs, intervertebral articulations, and the pubic symphasis.

Synovial Joints

In general, synovial joints can be described as freely moving joints. These are the most common joints in the human body. The main structural difference between synovial, cartilaginous, and fibrous joints is the existence of synovial capsules surrounding the articulating surfaces (figure 7-3). The synovial membrane is the innermost layer of the capsule and is important as it secretes a lubricating, synovial fluid, into the synovial cavity or space. The synovial space exists between the articular surfaces of the opposing bones that form the joint. The presence of lubricating fluid and joint space (distance between bones) contributes to joint mobility. Cartilaginous discs or linings may or may not be present.

We need to spend most of our time working towards an understanding of synovial joints as they have the largest range of motion and are directly involved in human ambulation and movement. Synovial joints can be named according to the type of architecture they form or their function, the most notable example of this is the "ball and socket" joint of the hip but there are several types of synovial joints.

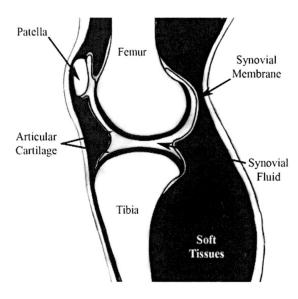

Figure 7-3. The basic features of a synovial joint.

Gliding joints. Sometimes called planar joints as they allow limited only gliding or sliding movements in one anatomical plane. An example this type of joint can be found between the carpals of the wrist and in between the tarsals of the foot.

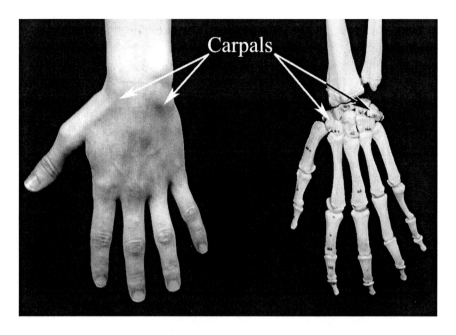

Figure 7-4. Gliding joints can be found between the carpal joints of the wrist.

Hinge joints. Hinge joints function exactly like the name indicates, like a door hinge. The elbow is a hinge joint and allows extension (open the door) and flexion (close the door) around one anatomical axis and in one anatomical plane.

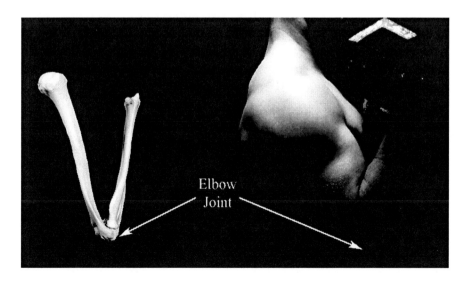

Figure 7-5. An example of a hinge joint moving around one anatomical axis is the elbow.

Pivot joints. As would be expected, rotation of a bone about another bone or bony feature is a characteristic of pivot joints. A good example of this type of joint is that between the atlas and axis bones of the neck (first and second cervical vertebrae). The atlas pivots on a bony projection coming off the superior surface of the axis and allows us to shake our head "no".

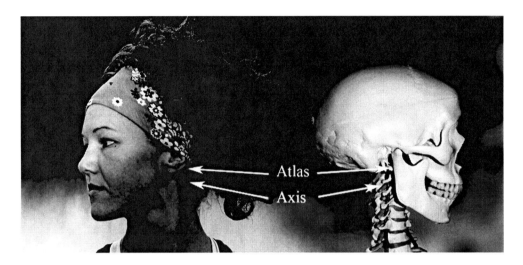

Figure 7-6. Pivot joints occur at several locations in the body. A visible example is found in the first two cervical vertebrae. The skull sits atop the atlas (first cervical vertebras). The atlas sits upon a bony protuberance of the axis called the dens. The skull and atlas pivot left and right around the dens.

Condyloid joints. A condyloid joint is where two bones, on with a concave surface and the other convex fit together with an odd shape. Example of this can be found where the wrist bones (carpals) articulate with the bones of the forearm (radius and ulna). These joints can allow movement around one anatomical axis as in the knee or in two anatomical axes as in the wrist.

Saddle joints. The joints are named such due to their resemblance to a saddle and a horse back (the actual articulation). The thumb, between the carpal and metacarpal is a nice example of a saddle joint.

Ball and socket joints. This type of joint allows for the greatest amount of movement in the all anatomical axes and planes. Think of a ball in a bowl to get the idea of the structure. The shoulder and hip joints are the most notable examples of ball and socket joints.

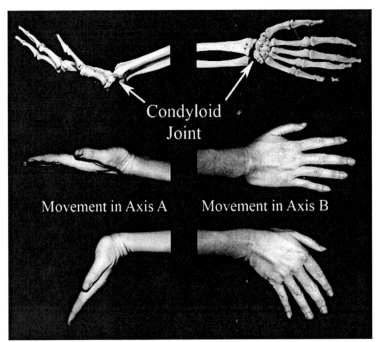

Condyloid Joint

Movement in Axis A   Movement in Axis B

Figure 7-7. An example of a condyloid joint between the bones of the forearm (radius and ulna) and the carpals of the wrist.

Figure 7-8. The saddle joint between the first metacarpal and a carpal bone, the trapezium, allows for movement around two axis thus provides humans one of their unique abilities, opposition of the thumb (right).

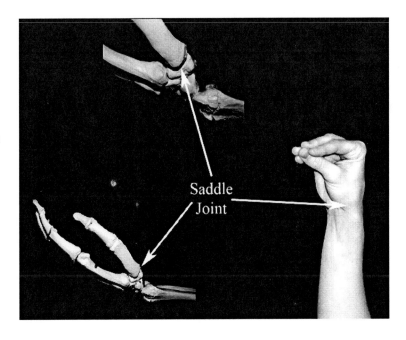

Saddle Joint

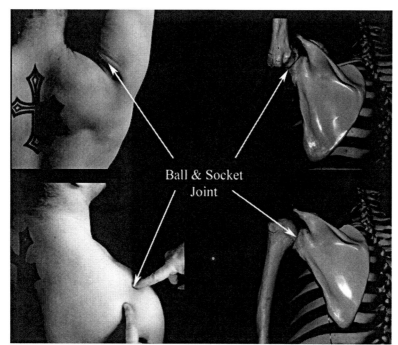

Ball & Socket Joint

Figure 7-9. The shoulder is an example of a ball and socket joint capable of moving in virtually any direction around multiple axes of movement.

There are some interesting applications of understanding joint anatomy and function. "Which way should that joint be moving?" is one interesting problem to solve. Innately when we watch a sporting event and somewhere on the playing field we see a joint that is bending the wrong direction we cringe, as we know that the bend is not supposed to be there. The sports medicine staff on the scene will evaluate the injury and by understanding the anatomy involved will be able to discern which anatomical structures are likely injured. This is an extreme but fairly common example seen in combative and plant-and-twist sports. On the other hand, as coaches or trainers, understanding how joints are built will assist in ensuring that complete range of motion for a joint is trained but not pushed beyond anatomical capacity. Pushing a joint beyond its structural limitations is called hyperextension (generically). You can also have hyperflexion, hyperadduction, and hyperabduction. Understanding that the olecranon (end of the ulna of the forearm) cannot be extended any further after being seated completely in the olecranon fossa (the elbow joint at full extension) will prevent a coach from telling a trainee to force further extension during exercise. This is another example of how a little knowledge of anatomy enhances safety of exercise and improves coaching ability.

8 - BUILT TO MOVE

One of the reasons I do what I do is simply because I wanted to know how to make myself a better athlete. From a young age (11) I read anatomy books. I read biology books. I read them not for fun but to look for answers to questions about improving my competition fitness in wrestling and weightlifting. Even in school I took elective classes that I thought would help me figure things out. In high school I took advanced biology where my senior research project was investigating the effect of different salt solutions and concentrations thereof on force production in isolated frog muscle preps. From my earliest recollections, I wanted to know how muscle was built and how it worked. I wanted to know how I could make things move.

To understand how things move we first need to take a look at how muscles are constructed. The first thing we need to know is that muscles are composed of thousands and thousands of individual muscle cells. Tiny muscles have a few thousand cells that can be less than a centimeter long. Massive muscles, like the latissimus dorsi that covers a huge portion of the back, have millions of muscle cells that can be up to 30 centimeters (about a foot) in length. All together muscle accounts for about 40% of total body weight in an average human. Let's dissect the muscle down to the cellular level, look at how a cell is built, identify its basic components, and briefly examine what each part does.

With his 1665 treatise, *Micrographia*, Antoni van Leeuwenhoek provided us with the first glimpse at the primary building block of living things – the cell. It took about 150 years of further study and experimentation by many scientists before enough evidence was acquired to allow zoologist Theodor Schwann to postulate in 1839 that "the elementary parts of all tissues are formed of cells". Schwann's works (along with the works of botanist Matthias Jakob Schleiden) led to modern cell theory, where the cell is considered to be the smallest structure having all the properties of living things. Those properties are:

- Homeostatic control, the ability to regulate the organism's internal environment
- Organismic composition based on one or more cells
- Metabolic activity - the consumption of energy through conversion of non-living materials into cellular components
- Capacity for growth
- Capacity for adaptation - the ability to alter form, function, or both over time in response to environmental challenge
- Responsiveness to external stimuli
- Capacity for reproduction - the ability to produce new organisms

BASIC CELL STRUCTURE

We are interested here in learning how the typical muscle cell is built. We will consider only the basic parts of the cell, as we want to understand the anatomy of the cell, not delve into the intricacies of molecular and cellular biology. Several early researchers identified common structures of all observed cells; the cell membrane, cytoplasm, and deoxyribonucleic acid (DNA) – although these terms were not used at the time (Figure 8-1).

Cell Membrane

Mammalian cells (you and I are mammals) have a bordering and constraining two-layer membrane made of phospholipids (phosphate containing fats/oils). The cell membrane contains the components of the cell; it is selectively permeable and allows some materials to pass into or out of the cell while excluding other materials from passage. All components of the cell are contained within the cell membrane. Anything the cell consumes or creates for export must pass through the cell membrane.

This semipermeable status of the cell membrane is eloquently dynamic. Each cell within an activated muscle orchestrates the passage of particular ions and the rejection of others in a predictable sequence during the contraction-relaxation cycle and is enabled to do so by the changing electrical potential across the cell membrane.

Cytoplasm

Inside the cell membrane is a complex collection of substances suspended or dissolved in water called cytoplasm. The cytoplasm is important as other sub-cellular structures are suspended in it. It is also important in that the first steps of cellular respiration (energy metabolism) take place in the cytoplasm. When discussing the cytoplasm of muscle cells the term sarcoplasm is frequently used.

DNA

All cells contain DNA, or genetic material. In the simplest of cells, DNA appears as a single loop floating free in the cytoplasm. In more complicated cells, like the mammalian cells that make up our body, numerous strands of DNA are encapsulated within a membrane-bound special structure called the nucleus. DNA is essential for life as DNA makes RNA (ribonucleic acid), which makes protein that makes function. In the most basic sense, DNA controls anatomy (how things are built) and physiology (how things work).

CELLULAR ORGANELLES

The invention of the oil-immersion microscope lens in 1870 led to a flurry of discovery in the late 19th century including the elucidation of other structures, or "organelles", that comprise the mammalian cell (Figure 8-1). Organelles are well-defined, large-scale structures (relative to the size of an individual cell) that carry out a specific set of functions within the cell.

Membrane Bound Organelles

Many organelles are "membrane bound", i.e. completely surrounded by a membrane. Membrane bound organelles are crucial as they allow different sets of biochemical reactions to be separated from each other so that they do not interfere with each other during simultaneous operation. A good analogy here would be a factory where different chemicals are kept in separate vats and mixed in separate and sequential mixing containers before a final product is produced. The different compounds and reactions involved in manufacturing the product are kept isolated from each other to keep the factory from producing a random mess of chemical goo. Compartmentalization of biochemical compounds and processes within membrane bound organelles prevents interference between different reaction pathways, provides the opportunity for sequential reaction control, and allows the cell to produce compartments with differing internal environments specific to each reaction's efficient completion.

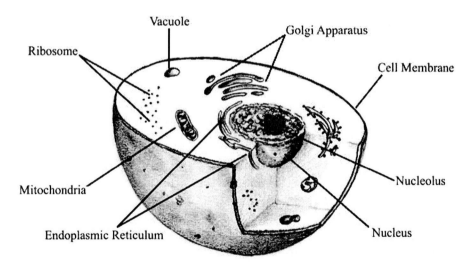

Figure 8-1. Schematic of mammalian cell organelles. The prototypical representation we see in our textbooks and above does not show the true reality of how organelles are distributed in the muscle cell. In muscle, nuclei are pressed up against the inside of the cell membrane and the other organelles are crammed between the contractile elements.

Endoplasmic Reticulum

The basic structure of the endoplasmic reticulum is an extensive membrane network of sac-like structures called cisternae. Like the plasma membrane, the endoplasmic reticulum membrane is composed of phospholipids and creates a bounded space, a networked lumen separate from the surrounding cytosol. There are three types;

- Rough endoplasmic reticulum (having associated ribosomes – see later description)
- Smooth endoplasmic reticulum (having no associate ribosomes)
- Sarcoplasmic reticulum (found in skeletal muscle and myocardium)

The *rough endoplasmic reticulum* plays a major role in producing lysosomal enzymes, secreted cellular proteins, and membrane proteins. It also participates in glycosylation (adding carbohydrate to proteins).

The *smooth endoplasmic reticulum* functions in many metabolic processes most notably the synthesis of lipids and metabolism of carbohydrates.

The *sarcoplasmic reticulum* is a variant found specifically in muscle. These variants differ in the composition of the proteins bound to their membranes and contained within their lumens. This difference in proteins present alters their respective functions. The smooth endoplasmic reticulum is a synthetic center and the sarcoplasmic reticulum is a regulatory center for calcium ion storage. The large stores of calcium within the sarcoplasmic reticulum are vital to muscle contraction and are rapidly released into the sarcoplasm which, in turn initiates contraction in muscle cells.

Golgi Apparatus

This organelle is composed of membrane-bound vesicles. Normally a few of these flattened sacs (between 5 and 8) will be in very close proximity but it has been observed that several dozen can be stacked in some instances in some cells. The Golgi apparatus takes vesicles from the endoplasmic reticulum, fuses with them, modifies the resulting contents before delivering them to their intended destination, which may include dumping the contents outside of the cell. They also assist in lipid transport in the cell and in creating lysosomes.

Mitochondria

A mitochondrion has a phospholipid bilayer membrane (outer and inner). The layers have different compositions (different lipids and embedded proteins present) and therefore have differing functions. There are five distinct compartments present within mitochondria:

- Outer mitochondrial membrane
- Inner mitochondrial membrane
- Inter-membrane space (between the outer and inner membranes)
- Cristae (the folding of the inner membrane)
- Matrix (area within the inner membrane)

Mitochondria numbers vary by location and cell type. A huge number of mitochondria are found in the liver where they can comprise up to about 20% of the total cell volume. They can also be found between the myofibrils (protein filaments) of muscle. They are often depicted as sausage shaped but their actual shape varies according to the how and where they are associated with cytoskeletal elements.

The most prominent function of mitochondria is rooted in energy metabolism. A set of reactions intimately involved in ATP production and known as the citric acid cycle (or Kreb's Cycle) and the electron transport system occurs within the mitochondria.

Vacuoles

These are membrane-bound compartments serving a variety of secretory, excretory, and storage functions. They may be called on to remove structural debris or waste from the cell, isolate harmful substances, store or release ionic molecules to maintain pH balance, along with other housekeeping functions.

Nucleus

The nucleus is the largest cellular organelle in mammalian cells and contains nearly all the cell's genetic material or DNA (the mitochondria contain some DNA). It has an average diameter somewhere between 11 to 22 micrometers (μm) and comprises about 10% of the typical cell's total volume.

The nucleus contains a viscous liquid, similar to cytoplasm, called nucleoplasm. Suspended in the nucleus is a "sub-organelle called the nucleolus, that is the site of ribonucleic acid synthesis and ribosomal assembly. Some cells are like red blood cells in that they are anucleate (no nuclei present). Others, like cardiac muscle cells, are mononucleate (one nucleus present) and some, like skeletal muscle cells, are multi-nucleate (many nuclei present).

Ribosomes

These small non-membranous organelles were discovered in 1955 after the invention of the electron microscope. Ribosomes, themselves built partially from ribonucleic acid (RNA), build proteins from genetic instructions passed from DNA to RNA. Ribosomes can be found "free", suspended in the cytoplasm, or they can also be bound to the endoplasmic reticulum, giving it the appearance of roughness and thus the name rough endoplasmic reticulum.

Lysosomes

These membrane bound organelles contain acid hydrolases (digestive enzymes) and work to digest worn-out organelles, food particles, or viral or bacterial pathogens that have been engulfed by the cell. The membrane of the lysosome allows a separation of the acidic lysosomal lumen and the neutral cytoplasmic environments.

SKELETAL MUSCLE

Muscle Cells

A skeletal muscle is made up of a bundle of *fascicles* bound together by a sheath of connective tissue. Within the fascicles are *muscle cells*, also called *myofibrils* or *myocytes*. They generally appear as long cylindrical cells with tapered ends—not the appearance of the typical mammalian cell. Their scale is fairly large, to the point of visibility with the naked eye (Figure 8-2).

The Myofilaments, Myosin and Actin

Bound within the muscle cells are long filaments of protein, myosin and actin, arranged in bundles. This is the business part of the muscle cell where contraction and relaxation takes place. Two sets of myofilaments, one thick the other thin are arranged in alternating lines with two identical sets across from them, separated by a space called the "*H zone*". Each set of myofilaments thus arranged is bound to another like set by a protein that makes up the *Z line*. This unit is known as a *sarcomere* (figure 8-2).

Thick and Thin.

Thick myofilaments are made primarily of the protein, myosin , and held in place, relative to other myofilaments, by titin filaments (another specialized protein). *Thin filaments* are composed of the protein actin held in place by another filamentous protein, nebulin, at the Z-Line (figure 8-3).

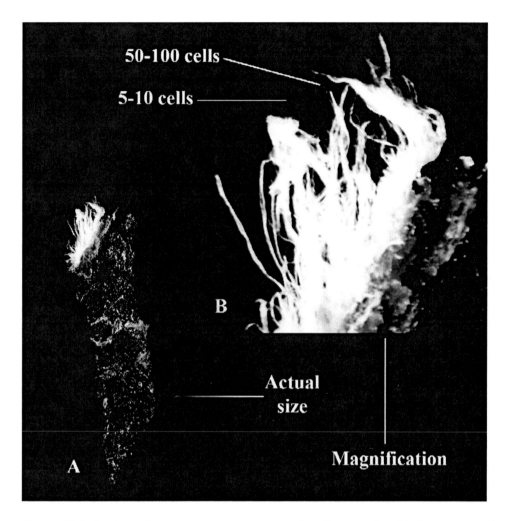

Figure 8-2. A trip to the grocery store or gas station for some beef jerky can provide some excellent perspective on muscle anatomy. Take a really dry piece of jerky and bite/tear it in half (A). Carefully look at the ripped end of the jerky. The frayed and cotton candy-like strands (B) are muscle cells or small groups of muscle cells.

Striated Muscle

This repeated arrangement of myofilaments, zones, and lines within the sarcomere gives the skeletal muscle its uniquely stripped appearance. Thus, it is called a *striated muscle*. Not all muscle cells have this characteristic banding pattern. Skeletal and cardiac muscles are striated; but smooth muscle cells from the vasculature, internal organs, and the muscles of the iris are non-striated and involuntary, controlled only by the central nervous system. Cardiac striated muscle is unique and of course, involuntary.

71

Troponin-Tropomyosin-Calcium Interaction

Actin has two other important proteins associated with its structure, troponin and tropomyosin. In the absence of intracellular calcium, when the muscle cell is at rest, troponin is attached to the protein, tropomyosin. In this bonded state the tropomyosin blocks the interaction of actin and myosin, preventing the muscle from contracting. But wait. Here comes an excitation wave! When the muscle cell is stimulated (depolarized), calcium is released from its stores in the sarcoplasmic reticulum. The free calcium ions attach to the troponin, negating its bond with tropomyosin and its block of the muscle contraction. Myosin and actin myofilaments slide toward each other into the H zone, shortening the muscle. When the muscle cells begin their return to a resting state (repolarization), calcium detaches from the troponin and returns to its home in the sarcoplasmic reticulum until the next stimulus arrives at the cell membrane. Tropomyosin once again binds to troponin and prevents contraction.

ORGANELLES AND DIFFERENTIATION OF MUSCLE FUNCTION

Organelles can be present in cells in varying numbers or absent completely, depending on the function of the cell. A red blood cell has no nucleus whereas a muscle cell has a huge number present. This tells us that a red blood cell is not meant to repair or recreate itself. It also tells us that a muscle cell possesses the capability of repair and growth owing to the presence of a large compliment of genetic materials in its nuclei. And indeed this is what happens. As a red blood cell ages or becomes damaged it is removed from circulation. A damaged muscle cell will use its genetic power to stimulate the production of repair proteins to re-establish normal function.

Fast Twitch and Slow Twitch Muscle Fibers

Another example of differential presence is in muscle cells. Most people are familiar with the concept of "fast twitch" and "slow twitch" muscle fibers. This is a simplistic concept but works well here for illustration.

Slow twitch fibers. There are large numbers of mitochondria in slow twitch fibers as oppsed to fast twitch fibers. This anatomical difference yields a functional difference in that the high numbers of mitochondria in slow twitch fibers are energetically efficient and thus fatigue resistant.

Fast twitch fibers. having greatly fewer mitochondria, fast twitch fibers fatigue within a few seconds of maximal contraction. They are also larger than slow twitch fibers and thus have more actin and myosin, meaning a larger force production capacity.

EFFECT OF EXERCISE ON CELL STRUCTURE

So it should be apparent, even with this short primer in muscle anatomy, that anatomical form dictates physiological function. This theme will be repeated over and over throughout your study of anatomy. It is very important to note that exercise training can change the anatomical structure of a cell. Whether it is a training induced alteration in the chemicals present or a wholesale architectural change in the cell's structure, the effects of exercise on the human begin at the cellular level before they become manifest in outward appearance or performance changes.

MYOFIBRILS AND MOVEMENT

Let's consider how the myofibrils characteristic of muscle cells function to produce movement – the physiology component of this lesson

The accepted mechanism of muscle contraction is a relatively modern concept, proposed in its most basic form in the 1950's by A.F. and H.E. Huxley. In *the sliding filament theory*, the two contractile proteins, actin and myosin, bind to each other intermittently and transiently when neurally stimulated to do so. It's a fairly easy concept to understand; your brain or a reflex mediated by the autonomic nervous system sends a signal out along a motor neuron (a nerve feeding information to a muscle). The neural signal hits the muscle cell, which triggers chemical events inside the cell. Those chemical events yield a binding of actin to myosin, energy gets expended, myosin changes shape causing myofibril and cell shortening and a production of force.

Think of actin as a ladder lying on the floor. At the distal end, on the last rung of the ladder, is something you want. Think of you sitting at the proximal end of the ladder (opposite end of what you want). How do you get the thing? Using your hands and arms you pull the far end of the ladder, hand-over-hand and rung by rung, closer to you until finally have the desired object within reach. In this analogy you have behaved like myosin, cycling your hand contacts and force generation against actin (the ladder) to accomplish movement.

Now think of doing this same task in tandem with someone else working with another ladder pulling their ladder and desired object in the opposite direction. In this orientation you and your partner are behaving like a sarcomere, the basic contractile unit of the muscle cell, pulling the ends of the system towards center. There are approximately 400 sarcomeres for every millimeter of myofibril. The structure of a sarcomere allows for a cumulative in-series shortening of the length of the entire muscle cell through an energy dependent mechanism (figure 8-3).

So the myofibrils, the long chains of sarcomeres that are contained within the muscle cell, generate the force. This force is transferred to adjacent cells by virtue of the connective tissue surrounding each cell. This connective tissue is called the endomysium and the endomysium surrounding one cell merges with that of its neighboring cells. There is then a secondary level of connective tissue structure that assists in force transference. Surrounding significant numbers (bundles) of muscle cells you will find a thicker connective tissue called the *perimysium*. The perimysium and all of the muscle cells within its boundaries is called a *fascicle*. A fascicle is easily visible to the naked eye (Figure 8-4). There is one more level of connective tissue structure that enables force transference, the *epimysium*. The epimysium is the connective tissue layer that bounds whole muscle (Figure 8-4). So each layer of this connective tissue tree is contiguous with the other and culminates in tendons at the muscle's sites of attachment to bone. The shortening of sarcomeres results in the shortening of the whole muscle and acts to bring the two bones attached to the muscle closer together, we have generated a muscle contraction. If a muscle shortens, it contracts.

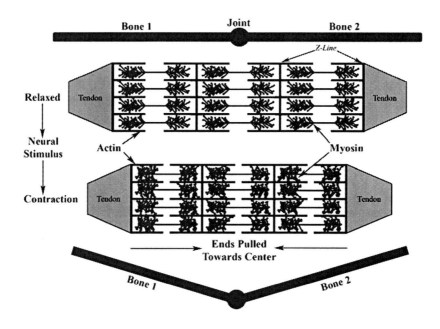

Figure 8-3. Sarcomeric anatomy and function. A sarcomere is bounded on each end by a Z-line into which actin is anchored. As the muscle contracts and brings the ends of the jointed bones together (bottom) the Z-lines are pulled towards center.

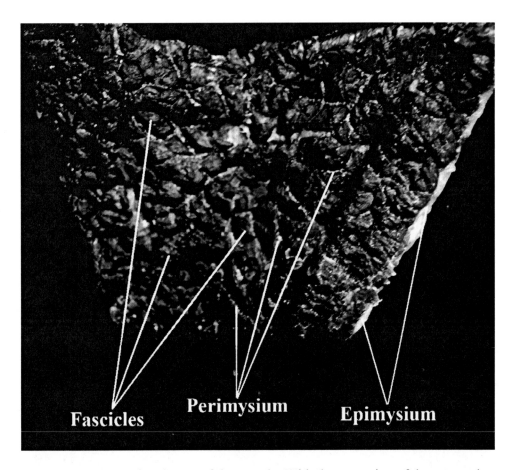

Fascicles   **Perimysium**   **Epimysium**

Figure 8-4. Connective tissues of the muscle. With the exception of the connective tissue surrounding individual muscle cells (the endomysium – not visible to the naked eye), a cross-cut piece of jerky or virtually any steak at the grocers meat counter can demonstrate the organization of connective tissues in the muscle. Bundles of muscle fibers are surrounded by connective tissue (the perimysium) and collectively called fascicles. Bundles of fascicles are bounded by more connective tissue (the epimysium) forming a complete muscle.

MUSCLE PENNATION AND FORCE GENERATION

A structural variation in the way muscle cells are situated within connective tissue affects the amount of force generated by the muscle. Pennation refers to the angle in which the muscle cells lie in relation to the long axis of the tendon on which they act (Figure 8-5). When a muscle's fibers run parallel to the long axis of its tendon it is capable of changing its length greatly and rapidly, although it produces moderate force (directly proportional to the force generated by sarcomeric shortening). A muscle of identical mass to the parallel muscle described above, but whose fibers insert into its tendon at 45° experiences amplification in force production. It's a math and physics thing with sines and cosines involved, but at the root of it there are more muscle fibers packed into the same space. The muscle will be capable of generating much more force than the parallel muscle although it will occur over a shorter range of motion and at a lower velocity.

How does this tidbit of anatomical and physiological information help us teach or program exercise? Think of all the calf exercise gimmicks out there proposing to improve vertical jump or sprint speed. Now think of the multi-headed gastrocnemius muscle (the big one of the calf). It has muscle fibers that are pennate to its tendon (the Achille's or calcaneal tendon) and as such is biased to produce force not velocity (you can't jump slow). Fiber type and mechanical lever issues further limit the contribution of the gastrocnemius to jump height. Placing a great deal of training attention to the gastrocs for the intent of increasing jumping performance is likely time misspent.

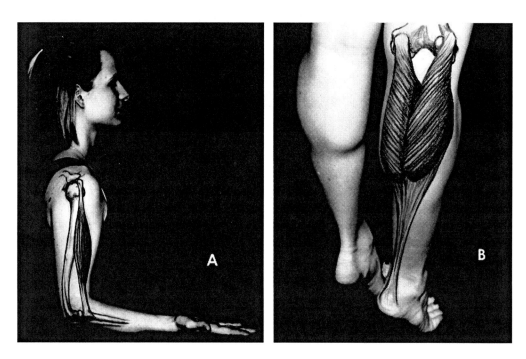

Figure 8-5. Muscle Pennation. The biceps brachii (A) are muscles with "parallel" muscle fiber arrangement (parallel to the long axis of the tendon). A bipennate muscle, such as the gastrocnemius (B) has different sets of muscle fiber angles radiating from a central tendon.

A conceptual understanding of the anatomy of muscle contraction makes it obvious that there is much going on within a muscle during contraction. And indeed there is, from individual molecules to the entire muscle. Muscle anatomy forms the structural basis of contraction. From the proteins that produce the force to the level of whole muscle action, a simple understanding of how things are built forms the core of our knowledge of how to change their structure to improve their function. The physiological processes inducing and supporting muscle contraction are numerous, diverse, and important to know. Making muscle move requires the involvement of the neural, cardiopulmonary, skeletal, and muscular systems making it an integrated physiological topic for later consideration.

"The last thing I want is a nurse or doctor working
on me who didn't buy their anatomy book."

- David Solimini

Isn't this a relevant sentiment for coaches, trainers, and PE teachers too?

9 – THE ANKLE & FOOT

Humans are built to move. They walk. They jog. They run. The basic structure of the human foot has not changed significantly for some four to five million years. It is an interesting design with many supportive and shock absorbing elements that make bipedal movement both possible and safe. We have supporting arches that carry the weight of the entire body and virtually any load placed upon it (figure 9-1). The many joints comprising the arches are quite well endowed with a multitude of muscles, tendons, and ligaments. The joints in the arches flex in order to dampen forces encountered when the foot contacts the earth underneath it during movement. The foot is a wonderfully engineered and purposeful anatomical feature, perfectly constructed to carry out its function. Da Vinci's renaissance depiction of foot anatomy, drawn from anatomical specimens, led him to state "The human foot is a masterpiece of engineering and a work of art." Due to its ability to support, the arch plays a central role in many of Da Vinci's architectural works.

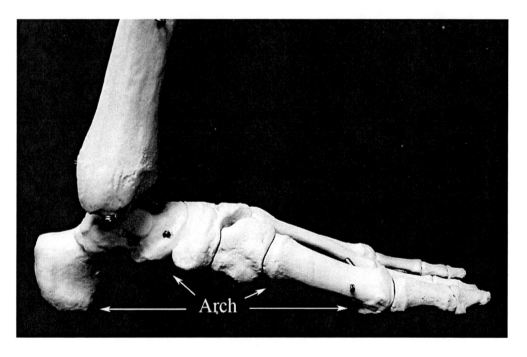

Figure 9-1. The longitudinal arch of the foot forms an effective supportive and shock absorbing structure. A transverse arch (running across the foot) performs a similar function. An analogous structure would be leaf springs in a car's suspension, quite robust.

BONES

The ankle joint (*talocrural joint*) above the foot is formed by the articulation of the *tibia* and *fibula* with the *talus* bone of the foot (figure 9-2). During standing and ambulation, the weight of the body is borne by the tibia and transferred through the foot; the fibula has no real anatomical basis of support in this system. This is similar to the wrist, where the ulna does not contribute substantially to the formation of the wrist joint. The distal ends of the tibia and fibula, where they articulate with the talus, can be located by palpating the *lateral* and *medial malleoli* (malleolus is singular) (figure 9-3). These two bony features are the hard bumps we usually point to as the ankle bones. They form the widest point of the ankle joint and are easily palpated. The talus sits in a groove formed underneath and within this structure.

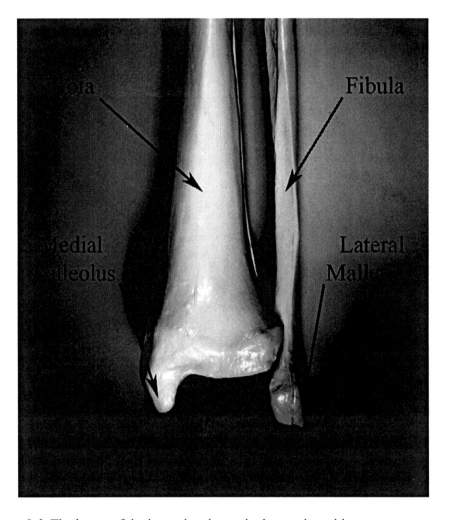

Figure 9-2. The bones of the lower leg that articulate at the ankle.

Tibia - This is the larger of the two bones of the shin or lower leg. It is bracketed by joints above and below - the ankle below and the knee above. Both ends of the bone are enlarged compared to the body or shaft. The distal end possesses a process, the medial malleolus, that extends lower than any other portion of the bone. The tibia is a weight bearing bone capable of withstanding forces at least five times bodyweight. While it is strong, it is one of the most frequently fractured bones in the body. One sub-type of fracture - stress fracture - is a condition where repetitive loading of the bone, with even light-to-moderate loads (distance running for example), overwhelms the bones capacity to remodel and repair itself. Over time, a line of architectural disruption, a hairline fracture appears. It is estimated that up to thirty percent of all runners will experience such a fracture during their training life.

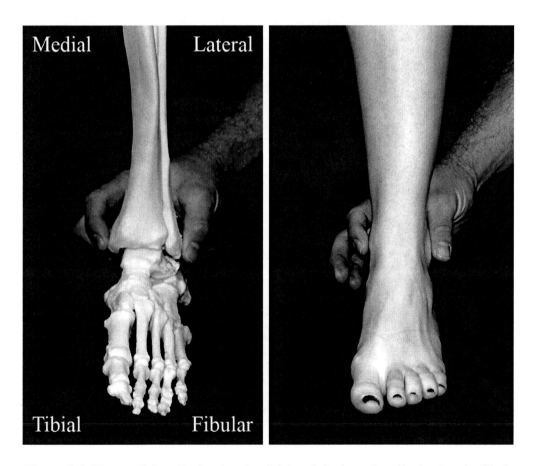

Figure 9-3. The medial malleolus (on the tibia) and the lateral malleolus (on the fibula are very easy to visualize and palpate.

Fibula - The fibula is the long and thin bone lateral to the tibia. It too spans the distance between the knee and the ankle joints. It has a small pointed expansion on its distal end called the lateral malleolus. When inspecting the ankle, the lateral malleolus occurs at a level along the leg lower than that of the medial malleolus.

The Tarsus

There are a total of seven irregularly shaped tarsal bones comprising the tarsus (figure 9-4). All seven occur between the distal tibia and fibula and the proximal metatarsals. These are important structures as they are essential architecture in the force dampening and supportive functions of the arch of the foot. These bones are the *talus, calcaneus, navicular, cuboid*, and the three *cuneiforms*.

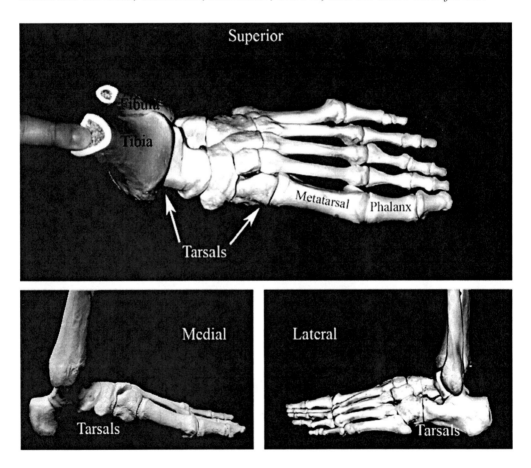

Figure 9-4. Medial, lateral, and superior views of the bones of the ankle and foot.

Calcaneus - The calcaneus is the largest tarsal bone of the foot and the easiest to locate. It forms the heel of the foot and articulates with the talus to form the *subtalar joint*. The superior and posterior surface can be located by palpating the attachment of the *achilles tendon* (calcaneal tendon). You can palpate the posterior, medial, lateral and inferior surfaces of the bone by walking your fingers from the posterior point of the heel forward.

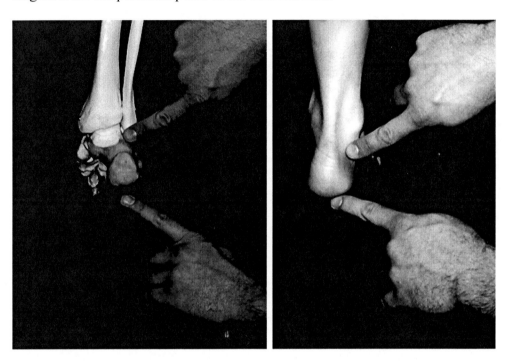

Figure 9-5. Identifying the calcaneus.

Talus - The talus lies between the tibia and the calcaneus. The superior articulation with the tibia forms the *talocrural joint* or ankle, and the inferior articulation with the calcaneus forms the *subtalar joint*. You can find the talus just distal to the tibia on the anterior ankle by pressing your fingers into the ankle (figure 9-6). Palpation can be aided by relaxing the ankle so that the tendons in close proximity do not interfere with palpation.

Navicular, Cuneiforms, and the Cuboid - The navicular articulates with the talus on the medial side of the foot. The cuboid articulates with the calcaneus on the lateral side. Three cuneiform bones (medial, intermediate, and lateral) articulate proximally with the navicular and cuboid bones and articulate distally with the medial four phalanges (toes). The cuneiforms form an arch, medial-to-lateral, across the foot known as the transverse arch. The longitudinal arch, which is formed by the articulation of the navicular and the talus bones, is found on the medial side of the foot and is the arch with which most of us are familiar (figure 9-1). The lateral side of the foot does not have much of an arch because the cuboid articulates with the calcaneus which in turn contacts the ground (figure 9-4).

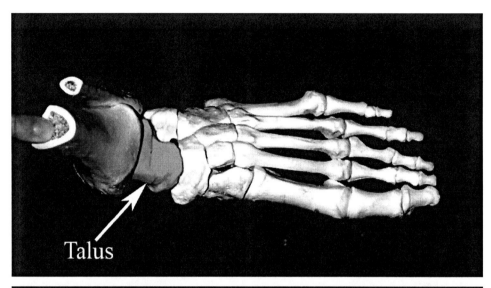

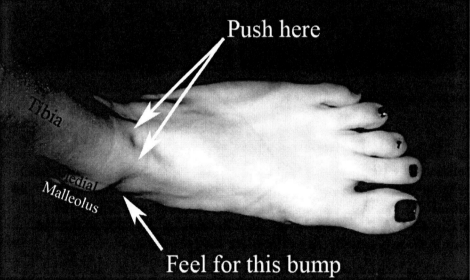

Figure 9-6. Locating the talus.

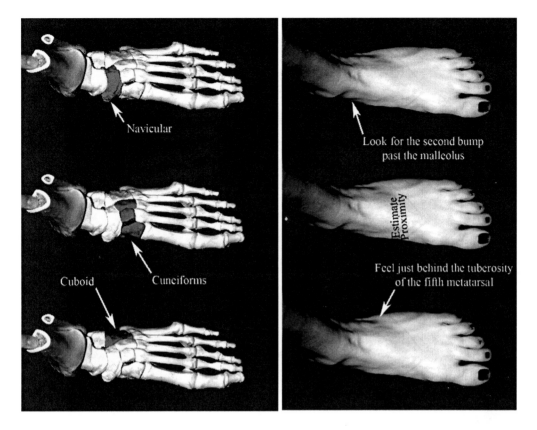

Figure 9-7. The anterior bones of the tarsus: navicular, lateral cuneiform, intermediate cuneiform, medial cuneiform, and cuboid. Identification of the navicular, cuneiforms, and cuboid may be difficult and you might need to use general spatial relationships to estimate their locations on a living human.

Metatarsus

Immediately anterior to the tarsus is a row of five relatively longer bones collectively called the metatarsus (figure 9-8). These are easy bones to remember as each is simply called a metatarsal. The nomenclature is by convention, the first through fifth metatarsals. The first metatarsal being the most medial (behind the big toe) and the fifth being the most lateral (behind the little toe). The metatarsals, specifically the second, third, and fourth contribute to the structure of the longitudinal arch and as their position is relatively higher than the flanking first and fifth metatarsals. They also contribute to the transverse arch in the forefoot. The metatarsal are maintained in close proximity by several and various short ligaments.

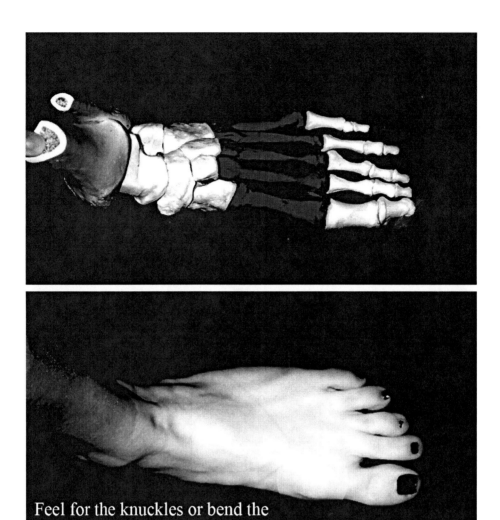

Feel for the knuckles or bend the
toes up to find the distal ends

Figure 9-8. The metatarsus.

Phalanges - In the foot the phalanges are the toes (or the digits) and like the metatarsals are referred to by number, medial to lateral (figure 9-9). Each phalange, with the exception of the first, is comprised of three small bones. Each of these is called a phalanx (singular = phalanx, plural = phalange) and is referred to by positional relationship to the body – proximal, intermediate, and distal. For example the second phalanx of your middle toe would be the intermediate phalanx of the third phalange. The exception to this convention is driven by the presence of only two phalanx in the first phalange (big toe, great toe or hallux - a name that will be useful later in learning muscle names). This results in only a proximal and distal phalanx being present. We will use this terminology again in the hand where the term phalange refers to a finger (or digit).

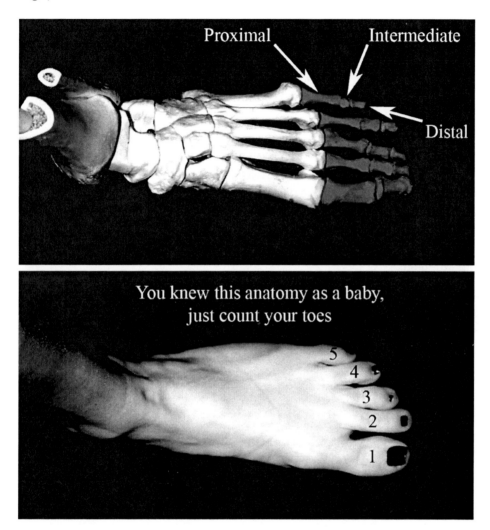

Figure 9-9. The phalanges and their component phalanx.

JOINTS

As mentioned earlier, the ankle is quite easily palpated. It is a prominent joint and its ligaments function as ligaments elsewhere in the body do – connecting bone to bone – and the major ones are fairly simple to identify. Conversely, the multitude of joints and ligaments of the foot, especially those of the tarsals are not easily visualized or palpated. We have to rely on illustrations to establish approximate locations for these joints and ligaments. Most commercial exercise professionals and coaches will not be called on to discriminate between the joints of the foot in practice, however clinical exercise professionals may.

Ankle Joint - The *talocrural* or ankle joint (refer to figure 9-4) is a synovial hinge joint comprised of two long bones, the tibia and the fibular and an underlying tarsal bone, the talus. The talocrural joint allows two foot movements with confusing names; dorsiflexion and plantar flexion (figure 9-10). *Dorsiflexion* is where you lift the forward portion of the foot up (rock back on your heels with the balls of your feet up). *Plantar flexion* is where you push the forward portion of the foot down (raise your heels off the ground as you go up on the balls of your feet). If we held to convention, dorsiflexion would simply be called flexion of the ankle and plantar flexion would simply be extension. But tradition trumps convention in this case. This an example of the necessity of using alternative terms to tell a trainee what to do. "Place your foot with the phalanges and anterior aspect of the metatarsals on this block. Now I want you to plantarflex the foot then control the descent into dorsiflexion" is not effective coaching language.

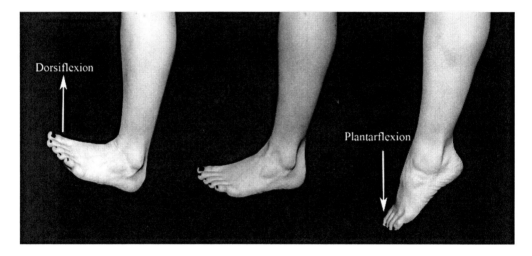

Figure 9-10. Dorsiflexion and plantarflexion of the talocrural joint.

The fibula, although not precisely an ankle bone, is held in close proximity to the tibia by the *interosseous membrane*. This sheet of connective tissue is a deep and non-palpable structure that transfers force between the tibia and fibula. This thin membrane has its fibers oriented in a downward oblique arrangement, from medial to lateral, and runs the entire length of the fibular and tibial shafts. This is one of the structures that may become irritated and inflamed with long runs using the heel-strike technique.

There are two major ligaments of the ankle, the *medial* and *lateral collateral ligaments* (figure 9-11). These two ligaments are sometimes called the tibial (medial) and fibular (lateral) collaterals. This latter convention makes their locations easier to remember. The medial collateral ligament is divided into three sections and has a third common naming convention, the *deltoid ligament* (remember the Greek delta shape). The ligaments function to limit side-to-side movement. The lateral ligament is divided into two sections and by virtue of the size of the ligament and the area of attachment, the lateral or fibular ligament is not as strong as the medial or tibial collateral ligament. This means that anatomically it is easier to injure the lateral ligaments.

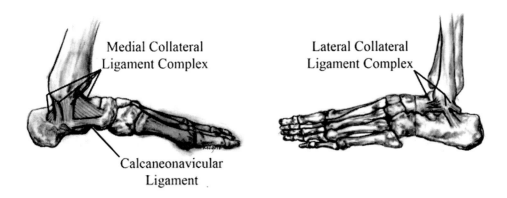

Figure 9-11. The medial collateral, lateral collateral, and calcaneonavicular ligaments of the ankle.

Subtalar Joint - The subtalar joint is also known as the talocalcaneal joint. It is a nonaxial sliding joint formed by the articulation of the convex surface of the talus and concave portion of the calcaneus that lies inferior and posterior to the talus. This joint facilitates side-to-side movements of the foot known as inversion (roll the sole of the foot to face medially) and eversion (roll the sole of the foot to face laterally) (figure 9-12). Again, the conventional terminology of adduction and abduction are not used in favor of the traditional terms inversion and eversion.

Intertarsal Joints - There are six other intertarsal articulations; talocalcanoeonavicular, calcaneocuboid, cuneonavicular, cuboidonavicular, intercuneiform, and cuneocuboid. These joints are simply named for various tarsals that articulate at the joint (figure 9-13).

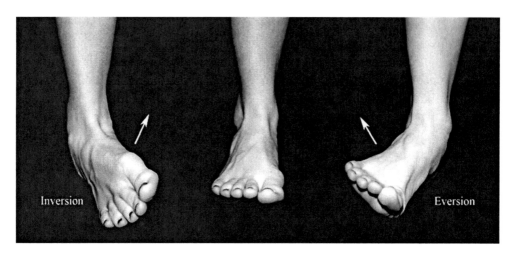

Figure 9-12. Inversion and eversion at the subtalar joint.

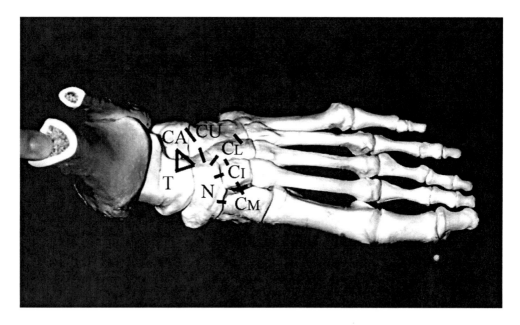

Figure 9-13. The intertarsal joints and their ligaments. When inspected in vitro, these ligaments envelop the tarsals and appear much more like rugged webbing than the lines presented here.

There are two ligaments of the tarsus that are palpable and are also important to the integrity and function of the arch of the foot. The longitudinal arch of the foot receives support from the simple interlocking of the tarsal bones, from the intertarsal ligaments, and it is also reinforced by the *long plantar ligament* and the *calcaneonavicular ligament* (figure 9-14). The long plantar ligament is attached to the midanterior, lateral, and inferior surface of the calcaneus and to the inferior tuberosity of the cubiod. The ligament then splits and extends to the proximal bases of the second, third, and fourth metatarsals. The calcaneonavicular ligament is medial to the long plantar ligament and attaches to the anterior calcaneus and the navicular. This is a broader but shorter ligament than the long plantar. Together these two ligaments are frequently called the spring ligament complex (remember the automotive leaf spring analogy for the arch). You can also imagine the structure here as the arch is a bow, the long plantar ligament as a long string attached near the ends of the bow and the calcaneonavicular as a shorter string with one end attached closer to the hand grip. When one of both of these ligaments fail, the ligament(s) lengthen and the arch becomes flatter. There is a variation in the height of the longitudinal arch with shallower arches termed "flat feet" or "pes planus." The *plantar aponeurosis* is a relatively thick band of connective tissue that also supports the arch of the foot. It attaches at the tuberosity of the calcaneus and runs forward to the heads of the metatarsal bones. It is fairly easy to palpate the plantar aponeurosis. Simply place the fingers across the bottom of the midfoot then alternately flex and extend the toes. Excessive impacts (as in many many miles of running) can irritate and inflame this structure, a painful condition called plantar fasciitis.

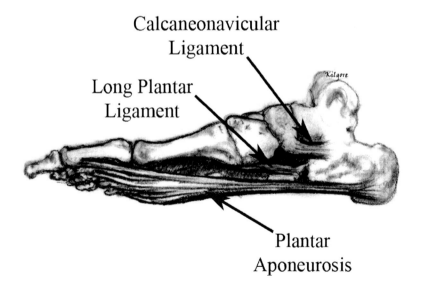

Figure 9-14. The long plantar ligament, the calcaneonavicular ligament, and the plantar aponeurosis.

Tarsometatarsal Joints - As one would expect, there are five tarsometatarsal joints (figure 9-15). These are all nonaxial sliding joints and display limited movement capacity. The first metatarsal articulates with the medial cuneiform and to a lesser degree the intermediate cuneiform. The second metatarsal articulates with all three cuneiform bones. The third metatarsal articulates with the lateral cuneiform. The fourth metatarsal articulates with the lateral cuneiform and the cuboid. The fifth metatarsal articulates with the cuboid. This set of joints are sometimes referred to as Lisfranc joints after the Napoleonic surgeon and gynecologist who first described their dislocation and associated metatarsal fracture in mounted soldiers who had fallen from their horses and had their foot trapped in the stirrup after the fall. In modern times it is occasionally represented in team sport where a player with his toes on the ground and heel elevated has the top of his foot struck by another player's downward moving heel.

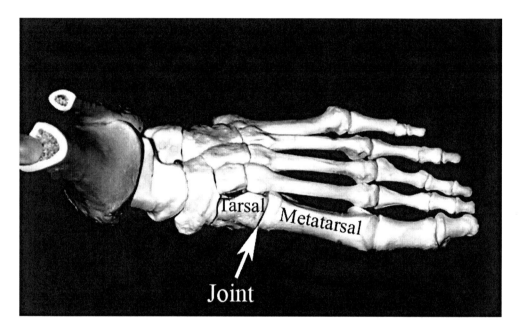

Figure 9-15. The tarsometatarsal joints. The first tarsometatarsal joint is indicated in the figure. All five metatarsals have these joints.

Metatarsophalangeal Joints - The metatarsophalangeal joints are formed by the articulation of the metatarsals and the most proximal phalanx in each phalange (figure 9-16). They are condyloid joints. The rounded surface of the metatarsals articulates with the saddle shaped shallow cavities on the proximal phalanx bones of the digits. The metatarsophalangeal joints have two collateral ligaments that course along the joint capsules and a plantar surface ligament. "Turf toe" is an injury (sprain) of the metatarsphalangeal joint at the hallux due to hyperflexion, hyperextension, hyperadduction, or hyperabduction caused by violent impact.

Interphalangeal Joints - Each of the phalanges have two interphalangeal joints. These are formed by the articulation of two adjacent phalanx (figure 9-16). The pollux is the exception as it has but one such joint owing to the presence of only two phalanx. These joints are easily identified visually by simple inspection of joint movement and then palpated. The interphalangeal joints, like the metatarsophalangeal joints, have two collaterals and one plantar surface ligament.

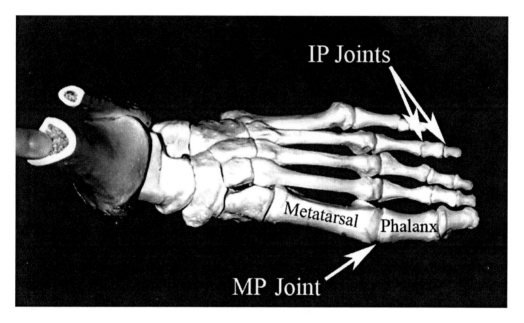

Figure 9-16. Metatarsophalangeal (MP) and interphalangeal (IP) joints.

Flexor and Extensor Retinaculum - An encircling band of connective tissue, the flexor retinaculum and extensor retinaculum, surrounds the ankle (figure 9-17). The simple function of this feature is to keep the underlying tendons and anatomical features in place for effective function. If this feature was not present the tendons of the ankle would bowstring out and away from the bones during contraction of their associated muscles.

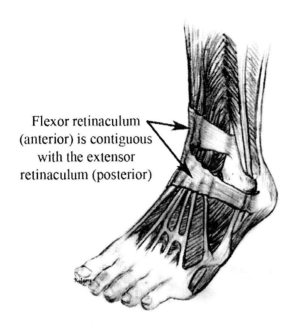

Flexor retinaculum
(anterior) is contiguous
with the extensor
retinaculum (posterior)

Figure 9-17. The
ankle retinaculi.

ANATOMICAL FUNCTION AND THE FOOT-SHOE INTERFACE

The foot has been a recipient of a considerable amount of attention in respect to the footwear we put on it during sport and exercise for about the past half century or so. A seemingly endless supply of innovations and techware is available to improve sport performance or reduce injury rate. This is not a real surprise as the athletic footwear market is extremely lucrative, with nearly 400,000,000 pairs sold annually to yield about $12 billion in annual sales. The stalwart Chuck E. Taylor shoe has sold on the order of 800,000,000 pairs worldwide since its introduction. Let's take an objective look at how the anatomy of the foot and the anatomy of the average athletic shoe affect each other.

When we consider the range of human activity and any possibility of anatomical predisposition to injury, one would assume that during the many millennia of human history there would be regular advances in technology that would either reduce injury rate or improve locomotive function - if there was indeed an environmental challenge and need. Strangely, such advances were not seen until the past century or so, a pitifully small segment of human history. Why was this? We consider shoe design to be a major component of exercise performance today, so why did our ancestors not do anything in respect to improving shoe design and function? And how do we know they didn't?

To ferret out more about this little niche of history, all we need to do is visit any well-inventoried art museum. Look at the depictions of humans at work, play, or war. In every era and locale of art history, you will see the human foot represented walking, running, lifting, and standing. The only modifications you will see to the bare foot depicted are thin sandals to protect the foot from sharp objects and you will see cloth or leather wraps to protect the foot from cold. A semblance of what we know as the shoe appeared no less than 5,300 years ago and certainly figuring out how to keep from cutting your foot or being able to prevent frostbite were amazing advances. But today we would perceive those as very low-tech solutions to simple problems. But from their first appearance, sandals and wrappings for the foot only varied in materials and construction method, not in elementary structure or purpose.

Figure 9-18. Pre-Christian Egyptian artifacts illustrate bare feet (as depicted here) or sandals as typical footwear. (Photographed at the Dallas Museum of Art)

Figure 9-19. Greek art of the 3rd century BC depicts bare feet (left) or sandals (right). (Photographed at the Dallas Museum of Art)

Figure 9-20. Mexican artifacts from approximately 100 BC predominantly show bare feet. (Photographed at the Dallas Museum of Art).

Figure 9-21. A Roman sarcophagus from the 2nd century AD shows a mounted soldier with even thickness sandals in battle. (Photographed at the Dallas Museum of Art)

Let's be critical here. If, as Da Vinci postulated, the foot is amazingly well suited to supporting and transferring force, why would we and why do we have super-duper air gel matrix torsion cushiony shoes that are touted to be the pinnacle of running performance footwear? Why try to improve upon something already well suited to its function, recreate the wheel so to speak? The evolution of athletic footwear is quite troubling in terms of solving, or at worst, producing exercise technical problems. Our ancestors functioned quite well in minimalist footwear or with none at all. Has the human condition changed so much for us to "need" advanced insoles, cushions, and even one particular structural shoe element we take for granted, the elevated heel?

Let's tackle this issue in two parts; (1) the evolution of the heel and (2) the evolution of shoe cushioning. Ideally we should find that the addition of the heel and shoe cushioning solved some identified problem in movement.

Heel History

As we have noted already, the shoe heel does not appear in art works of antiquity in sport, war, or daily life. Unsubstantiated or even substantiated stories about where the heel came from are sparse. There are three plausible explanations circulating. The first would be that the Roman's added height to the rear sole of their soldier's footwear to increase stride-length thus enabling them to cover more distance with the same number of strides. That sounds like something a military organization might actually do, but actual records or artistic depictions of this are not present in available archives (see figure 9-21). And why would they elevate only the heel? A thicker shoe sole along the entire length of the shoe would have accomplished the same purpose and would be easier to produce.

A second possible origination for shoe heels is attributed to the Hungarian Hussars (mounted military troops) somewhere during the 15th century (compare figure 9-21 to figure 9-22). It has been suggested that the heeled boots they sported were designed specifically to add stability and control to the foot-stirrup interface. An astute student with a cowboy background once surmised that the Hussar heeled boot would also enable the rider to "kick the crap out of" foot soldiers. The art of the era is replete with depictions of Hussars in heeled boots.

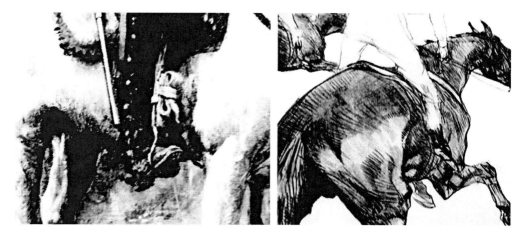

Figure 9-22. The boots worn by the Hussars influenced calvaries throughout Europe (left) and heeled footwear remained common following their introduction. Equestrian events, such as horse racing used, and continue to use low-heeled boots as depicted in 1899 (right). (Photographed at the Dallas Museum of Art).

The last commonly espoused origin of the heel traces back to Catherine de Medici in the mid-1500s. It has been suggested that she was sensitive of her diminutive stature and used elevated heels to boost her physical presence. There is some evidence that heeled shoes existed in Italy in the years prior to Medici, but she is believed to be the popularizing factor in their wear by nobles and the aristocracy. This is a trend in footwear (and fashion in general) we will see repeated throughout history, the general population are swayed more by elite endorsement than actual function. There are museums devoted to the history of heeled shoe fashions from this point in history to modernity (Bata Shoe Museum in Montreal for example).

So we have three basic possible reasons for the development of the heel; (1) increased stride length, (2) stirrup control, and (3) vanity. So in terms of running or human movement, the highly tenuous attribution to using the elevated heel to increase the distance an army can cover by the Romans is the only performance enhancing explanation for the heel. Getting marching troops from point A to point B faster than your enemy is important. But if this provided a significant tactical advantage, would we have not seen other military forces adopting the use of the heel in their military footwear? We don't see evidence of this in pictorial references from other contemporary or subsequent civilizations influenced by the Romans.

The other two historical possibilities, although documented, do not provide for a benefit to human walking or running performance.

So how did we end up with heels on running shoes? And do they actually do anything to improve performance or do they do something dastardly, like introduce an artificial running environment facilitating the introduction of a non-native running technique?

Do an experiment. Take your shoes off and go run down the street (or other HARD surface). Run fast and run slow. How do you run? If you are normal, you will probably contact the ground with the ball of your foot first with every stride. You do this intuitively without thinking because your body is doing what it is designed to do. In this case, the arches of the feet are absorbing the impact forces allowing you to run safely, comfortably, and without injury. If you are on pavement, you will have a hard time convincing your body to allow you to run with a heel strike first with bare feet, it is very jarring. You can do the same exercise by running in place – your body will not let you do a heel strike unless you force it. So here is the takeaway point, it is natural to have a ball-of-foot strike and not so natural to have a heel strike when running in bare feet. If we were destined to be more robust and efficient with a heel strike, would we not have evolved to run in that manner intuitively and under all conditions, especially those to which we have been exposed throughout human history?

Let's think back to the Romans. If an elevated heel (or sole) allowed a marching army to cover more distance, would this idea extend to running shoes? Would a shoe with a heel allow a runner to cover the same distance with fewer strides than a thin flat sole (or no shoe)? It's time for another experiment. Put on your best and most fluffy running shoes. Take one step forward to a heel down and toe up position (leave your trailing leg behind). Note the distance your foot-contact is away from your body. Now slowly point your toe down until the ball of the foot is in contact with the ground and your heel is slightly elevated. How far is the point of contact away from your body now? An interesting observation no? It looks as though a ball-of-foot strike adds more distance to the stride length than a strike with an elevated heel. So assuming our turn-over rate is the same between conditions, it appears we will run faster with a ball-of-foot strike than a heel strike with an elevated heel. So in attaining maximal running velocity, the presence of a heel is a non-contributor. If you want to verify this, just look at the footwear of sprinting athletes and do a video analysis of their foot strikes, fast runners do not used heeled running shoes and do not heel strike.

But there is more to it than this and we can't just dismiss the heel's possible benefits before examining a couple issues. The first is that a marching army is marching, not running as its mode of movement. So a thicker heel just might achieve a step-reducing objective effectively over long distances. The added thickness would also likely have extended the life of the shoe as well. Durable military equipment is a necessity. There may be a similar benefit with running shoes in the present. Most consumers of elevated heel running shoes are not racers, or even recreational runners interested more in achieving target heart rate than winning a race. They are the average persons of modernity. Those who spend the majority of their ambulatory life walking, and they will do so in the heel-toe walking gait we develop naturally in childhood. The possibility that an elevated heel will get us from point A to B a miniscule number of seconds faster during our workday is of little concern. We are by and large a sedentary society and do not cover significant distances other than by car. So since the heel provides no viable performance advantage, why have them? We need to consider an alternative viewpoint. Added thickness of the heel may be useful in another way.

Cushioning History

Could the value of the modern running shoe lie in the cushion factor? Shoe cushioning and cushioned supports are an invention of modernity. The evolution of what we know as the running shoe, cross-trainer, tennis shoe, sneaker, trainer, and a variety of other names, was made possible by Charles Goodyear's rubber vulcanization process. Patented in 1844 and widely used within a decade, the process made rubber heat and cold resistant (boat shoes constructed from native rubber in the 1820's failed to tolerate environmental demands). Goodyear's process allowed rubber's unique characteristics to be exploited year round in a variety of applications. By the late 1800's shoes were being manufactured from canvas (or kangaroo skin) and vulcanized rubber. These early shoes were not intended for performance enhancement, or athletics at all, rather they are an attempt to produce an inexpensive shoe to market to the public. Both US Rubber (Goodyear) and Colchester Rubber Companies produced these shoes before 1900.

These early rubber shoes, gum-shoes, sneakers, or plimsolls were rapidly noticed by athletes and coaches in sports played on hard surfaces. Their ability to reduce slipping was quite useful to performance more so than their hard-soled predecessors. The Spalding Company introduced a basketball shoe in 1907. By 1917, US Rubber was marketing a rubber "tennis shoe" and the Converse Shoe Company had introduced its basketball shoe, the All-Star (re-named in 1923 to Chuck Taylor All-Stars). All of these early sports shoes shared similar construction with essentially flat soles and insoles and canvas tops. The guts of athletic shoes for the general public would now remain static for about 25 years, save for the incorporation of new synthetic materials and superficial design elements for marketing purposes.

Although the appearance of supports and cushioning in shoes occurred at essentially the same time as rubber shoes developed, they are independent events and do not share a common history until very recent decades. The first supportive or cushioning element for shoes is widely attributed to William Riley who, in 1906, developed an arch support for wait staff and other workers who spent long hours on their feet. This arch support company would later evolve into the New Balance shoe company in the second half of the century. The use of rubber as a cushioning element in work shoes actually was elaborated in Butterfield's 1900 patent application for shock absorbing work boots. And the most famous proponent of cushioning and support, Dr. Scholl, patented his first arch support in 1911.

The first intersection of the worlds of cushioned support and sports shoes occurred in the 1930's when US Rubber released their shock-proof arch cushion in their Keds line of shoes and Germany's Adidas included arch support in their athletic shoes. It is prudent to note here that Keds were not intended as uniquely sports shoes, they were shoes for every purpose. Even with the arch support in place, the heel of these shoes, and all running shoes, remained at virtually the same level as the forefoot, a design that had not changed for millenia.

In 1962 New Balance introduced the Trackster running shoe. It had a unique rippled sole, a mild wedge heel, and arch support. In the same year Phil Knight and Bill Bowerman partner to form Blue Ribbon Sports, Inc. the predecessor to the Nike Corporation. In 1972 Nike released the Waffle Racer with waffled soles, a mild wedged heel, and a cushioned midsole. Since that time, the race has been on so to speak, in the development of cushioning elements for running shoes. Terasaki patented an air-cushioning device in 1972 and since then there have been hundreds and hundreds of cushioning and force-dampening gimmicks included in the running shoes marketed to the exercising public – under the premise of safety, comfort, and performance.

Now here we get to a controversial part. Even though athletic shoe companies have an ever growing menu of shoe features and shoe models, the shoes that win running races, sprints and up, have not changed much in concept or construction since the early 20th century. Look at Nike's newest Zoom Miler and compare it to the Adidas shoe worn by Emil Zapotek in 1948. You will see similar structure; minimal heel, some support and cushioning, lightweight upper, laces, and that's about it (figure 9-23).

So why no super cushioning and big fluffy heels in competitive running shoes? Easy. Performance. Can you jump higher barefooted or with big marshmallows on the bottom of your feet? Bare-footed of course. The less material and the less compressible it is in between your feet and the earth, the better we transfer force to move our bodies. A track flat does not absorb or dissipate the force you are generating with the muscles that move you. A track flat gets you out of the blocks faster, helps accelerate you faster, and maintain running velocity. Now this is assuming you run with good running technique. And this is a BIG assumption. How many people have actually been taught how to run? If your learning experience in running for athletics is similar to mine, it was a football coach telling you to go run around the field or a wrestling coach telling you to go run down to the highway and back at the end of practice. No one, ever, in PE or in sport actually took the time (or more likely knew how) to teach efficient and safe running technique. No one ever bothered to correct heel-strike running technique in favor of faster, more efficient and shock-absorbing ball-of-the-foot strike technique (not to mention the plethora of other flaws).

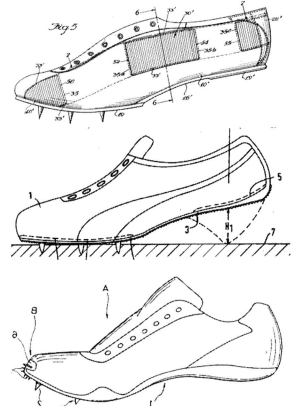

Figure 9-23. Comparison of a track shoe from 1951 (Top – from patent by Shapiro), a Puma track shoe from the seventies (Middle – from 1978 patent by Dassler), and a Mizuno track shoe from 1998 (Bottom – from patent by Kaneko).

Probably right now you are questioning all of this because shoe design is purported to be highly scientific. All of the highly engineered modern sports shoes are supposed to be safer and perform better aren't they? Well it's sort of a quagmire. If you run with a heel strike technique, a technique that maximizes repetitive shock to the body, yes, those cushy heels and marshmallow insoles will dampen the forces experienced and maybe fend off a case of shin splints or a stress fracture or two. In fact in Blue Ribbon Sport's (a.k.a., Nike) 1976 patent application, reduction of injury for runners who heel strike during running, not improve performance was the design goal (most performance improvement claims for running shoes has been focused in the development of track spike technology). Cushioning for heel strikers is the ***pro*** of high tech cushioned running shoes. The ***con*** is they can actually facilitate running with poor technique by virtue of allowing the runner to bypass his innate shock absorptive elements, the arches of his feet, and land on the heel. When you use a heel strike technique, the calcaneous (heel bone), is abruptly loaded with the force that would have been dissipated if the force would have been passed through the arches. Defeating the body's protective anatomy is risky. This also causes a transient deceleration of the body (so much for running fast). And unknown to most, there is another risk added. If you wear big cushy shoes, the chance of rolling an ankle is higher than if you wear flats. Marshmallow feet change proprioception, balance, and foot stability. You can see this in any individual wearing heavily cushioned shoes. They minutely sway back and forth as their postural reflexes constantly search to find a center of balance. Add movement into the mix and the body is hard pressed to find a constant and repetitive center of balance, the one needed for consistent technique of any kind.

So by wearing cushioned shoes are we reducing one type of injury at the expense of increasing another? Should we wear cushioned shoes with elevated heels or should we wear flatter shoes? These are hard questions to answer. But given the misguided conventional wisdom that long slow distance is the best way to get "fit" coupled with the normal human tendency towards "if a little is good, more is better", people will run way too many miles, way too slow, way too often to actually improve fitness. In such a case, especially if no *experienced and competent running coach* is available, then spending the money on cushion tech is probably wise.

Here is some more heresy. Learning correct running technique is more important than your shoes. How would I describe appropriate running technique? Remember arches are built to support our weight during all ambulation. Spend some money on learning how to run on the balls of your feet. If you are a fitness professional, learn how to teach running. Learn how to correct bad technique through apprenticing with EXPERT running coaches or attending their seminars (the most widely available seminars are POSE, CrossFit Endurance, and ChiRunning). Learn how to teach a variety of running events by attending a USA Track & Field coaching course.

All those conventional wisdom and on-line gurus out there will probably say "look at all those endurance runners who heel strike, they run fast and that means they can't be wrong." They will also probably say that "Running on your toes or balls of the feet for sprinters is fine but you can't keep running on the balls of your feet for a long distance or a long time." And they will point to the Boston Marathon racers wearing well-padded shoes as a point of support of this – that you need heavily cushioned shoes. OK. Let's consider long distance runners. Like any athlete who competes in varied environments, these athletes will have multiple shoes for racing in specific conditions and for training (figure 9-24). A competitive marathon is a very specific environment that produces profound fatigue, and with fatigue exercise technique decays. I maintain that a **racer** will be able to use correct running technique throughout a marathon with correct preparatory training using minimal heel cushioning. However, a **runner** will likely not be so well prepared and the big cushy shoes will likely be a godsend of comfort. For clarity we are using the Hunter S. Thompson definition of a racer and a runner from his description of Honolulu Marathon participants (*The Curse of Lono*, 1983):

> "The racers run smoothly, with a fine-tuned stride like a Wankel rotary engine. No wasted energy, no fighting the street or bouncing along like a jogger. These people flow, and they flow very fast.
>
> The runners are different. Very few of them flow, and not many run fast. And the slower they are, the more noise they make. By the time the four-digit numbers came by, the sound of the race was disturbingly loud and disorganized. The smooth rolling hiss of the racers had degenerated into a hell broth of slapping and pounding feet"

Racers are in contention to win. Runners are in contention to finish.

But in reality, how many of the millions of recreational runners actually run marathons? Not a large percent. So the vast majority of recreational runners will run just a few miles in a session and can adapt their musculature to accommodate correct forefoot strike technique throughout the exercise session and will be better served with little heel and little cushion. Still skeptical? Let's do another experiment. Get up, take your shoes off, and start running in place. Which gives out first, your cardiorespiratory endurance or your calves? (You'll probably get bored before either happen, but I hope you get the point) Even if your calf muscles start to ache, guess what? They can be trained to last longer, they will adapt just like any other muscle used in endurance running. Fluffy shoes provide a crutch, an easy out, an ability to develop poor technique, especially in those who have not been coached – and maybe at times in some that have been coached. You can run safely in running flats, but like any other exercise activity there is a need for common sense and progression. Wholesale changes in technique and equipment require time and titration to facilitate learning and safety.

And now for the obvious, a cushy pair of high tech running shoes won't make you run faster. Intuitively you "know" this already. How many of you with kids go out and buy them a new pair of running shoes and as soon as they get them home and on their feet, they run around like crazy and ask you "am I faster?" What do you tell them? High-tech and force dampening shoes do not make you go faster, good technique and appropriate conditioning of the human body do ... add a pair of flats and off you go to the races.

Figure 9-24. Three shoes for three different purposes. These shoes, used by a NCAA distance runner, are for shorter cross-country racing (top), longer hard surface racing (middle), and training (bottom). This particular runner contacts the ground first with the ball of the foot during racing and interchangeably with the ball of the foot or heel during training depending on the tempo of the training run. There seems to be a relationship between duration of run and cushioning. Shorter cross-country races have the least cushion and lowest heel. Longer road races use a minimal heel cushion. And since runners love to put in the miles, the trainers experience the highest single session mileage and are the most extensively padded and have the largest sole area for distribution of impact. Heel strike from fatigue laden miles, or is it just using inefficient technique because the shoes allow it?

The Other Important Pair of Shoes

Walkers, joggers, and runners may comprise the largest component of all recreational exercise participants, but we cannot neglect to address the other major fitness activity associated with getting fit, weight training.

Ever thought about the shoes that you wear to the weight room? Of course you have. You've actually spent some time thinking about which shoes to wear, and you probably have a pair designated as your "gym shoes". How did those shoes earn that illustrious title and serve such a noble purpose? Suitability for the task? Performance enhancement? Safety? Ideally, but not usually. Comfort and looks, just as in running shoe selection, seem to be the main criteria associated with gym shoe choice. This is a problem if your training includes any free weights at all. Most of us would never consider wearing a pair of Bruno Magli's (look it up) to play racquetball. They are built to look good with your Valentino suit, not to perform well on the court. While this may be obvious to some, many of us will make an equally poor footwear decision and wear running shoes to the gym to lift weights.

Proper footwear in the weight room is important, especially if you are lifting free weights. When we lift weights we want two things to happen; (1) all the force our body produces under the bar should contribute to moving the weight and (2) the weight needs to be controlled in a safe manner. If we lift in a running shoe, the ones we talked about previously, it's akin to trying to lift while standing on a giant marshmallow. The soles of the running shoes, the marshmallow, will absorb and dissipate a large amount of the force generated against the floor that should be directed towards moving the weight. A gel or air cell shoe is a OK for reducing the impact shock that causes the repetitive use injuries associated with the decision to run with a heel-strike running technique. But in the weight room, shoes should provide for the efficient transmission of power between the bar and the ground. You can't lift as much weight in the wrong shoes.

The second issue is control of the weight – and your body – while standing on an unstable surface. A compressible medium placed between the feet and the ground will behave inconsistently enough during each rep to alter the pattern of force transmission every time (figure 9-25). This means that the subtle points of consistent good technique on any standing exercise are impossible to control. And there is an increased chance for a balance or stability loss-induced injury while lifting heavy weights, since perfect balance cannot be assured on an imperfect surface (this concept extends to any weighted exercise on exercise balls – other than rehab purposes these faddish gimmicks have limited utility).

Weightlifters and powerlifters have known this for more than 50 years, although the shoe choices available for their purposes were formerly quite limited. Until the 1970's, combat boots, Chuck Taylor's, wrestling shoes, patent leather oxfords (see old photos of Paul Anderson), and even dance shoes (see old photos of Vince Anello) were the shoes used for lifting heavy weights.

To be stable and perform optimally, a weightlifting shoe needs to be snug fitting, provide exceptional support, and have a non-compressible wedge or modified-wedge sole with neoprene or crepe for traction against the floor. Many will lace all the way down to the toe for adjustment to individual foot width, and will have an adjustable strap(s) across the metatarsal area for added lateral stability. When Adidas from Germany and Kahru of Finland became available on a limited basis in the US, weightlifters finally had the opportunity to use equipment specifically designed for their activity. High topped and not especially stylish, these shoes had minimal appeal to the fashion conscious, but lifters loved them because they *worked*. (Obscure Factoids: Adidas actually purchased the three-stripe trademark from Karhu. And did you know that the three stripes were introduced to all of their shoe models to add metatarsal support?)

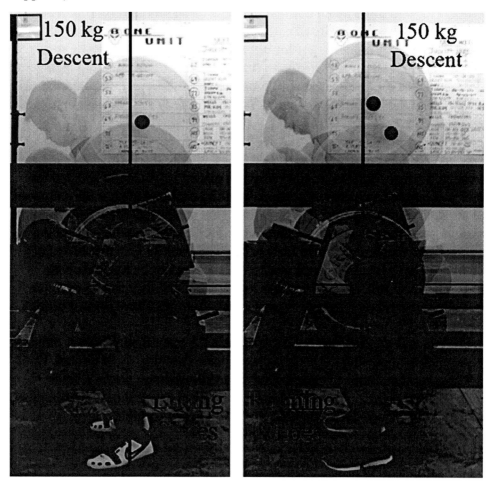

Figure 9-25. Notice how far backwards the bar oscillates when wearing muffin soled shoes. The black dots mark the end of the bar on it's descent to the bottom.

But there was a scheduling problem relating to the introduction of weightlifting shoes to the US: the gym and fitness club industry had just been revolutionized by the simultaneously-evolving exercise machine industry. Having removed the factors of balance, coordination, and technique from the equation, exercise machines temporarily sidelined the development and availability of weight training shoes. Over the past two decades, free weights and the benefits of their use have crept back into gyms and fitness clubs everywhere. The need for weightlifting shoes re-emerged without a supply beyond the stalwart Adidas corporation's Power Perfect, Equipment, and Adistar models. Other major shoe brands like Nike, Puma, and Reebok began to experiment with weightlifting shoes. A number of foreign brands such as Do Win (China), Rogue, and Power Firm (Canada), as well as several others have also competed for a share of the growing US market. All these companies offer shoes that are designed for competitive weightlifting or powerlifting, but that are good for all basic lifts, given their exemplary support and incompressible heel design. A variety of powerlifting shoes with essentially flat soles and no heel lift, much like track flats or wrestling shoes, are also available from powerlifting equipment houses like Inzer (USA), and also work for basic exercise purposes.

How about the heels in Olympic lifting shoes? What are they for? Shoulder control. Strange but true (remember body geometry affects exercise technique). A higher heel allows the shin angle to become more parallel to the floor, shifting the hips closer to the heels, thus allowing the torso to be more upright in the bottom of a squat without significant vertebral deformation (figure 9-26). This is a relatively nice perk when doing a snatch (one of the Olympic events). The upright torso enables the lifter to use more shoulder musculature to better control the barbell he is holding overhead, more so than the same lifter would have in flatter shoes in a more forward leaning torso.

Most die-hard lifters will have a favorite shoe to lift in and they will not train without the correct shoe. They might have a squat shoe and a deadlift shoe. Or they might do everything in their Olympic lifting shoes. Their experience should tell us all something.

Another pair of shoes to buy? Is it really worth it? Yes. Effective training yields superior results. Safe training yields fewer training injuries. The logic is inescapable. For as little as $40 for a pair of old-school Chuck Taylor's or as much as $200 or so for the state of the art Adidas shoe, you can have the right shoe for the right job. The right shoe is important for performance and safety, and for as little as half the cost of a premium squishy running shoe, you can look and lift like a pro (figure 9-27).

Figure 9-26. The effect of heel height on snatch bottom position - 56 kilogram lifter snatching 80 kilograms. The track flats on the left require a more vertical shin position, a less than neutral low back arch (slight rounding), and a head-through position. The higher heeled lifting shoes on the right allow the knees to move forward, the hips to move under the bar, and the back to be in a more neutral anatomical position. These photos of the same lifter were taken seven months apart at national events.

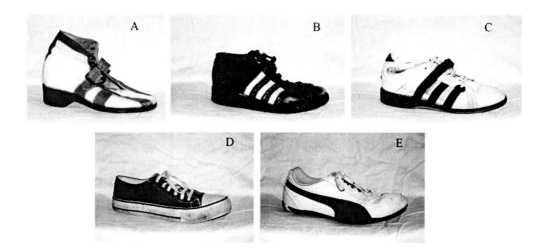

Figure 9-27. Modern and not so modern lifting shoes from the authors collection (A, B, C) and shoes used for lifting and other exercise activities when combined in the same workout (D, E). A = Korean lifting shoe from 1973. B = Adidas lifting shoe from 1977 (my dog chewed the metatarsal strap off ... really). C = Adidas lifting shoe from 1999. D = Chuck Taylor All-Stars. E = Puma shoes with no heel and relatively stiff soles.

MUSCLES

The skeletal architecture of the foot is essential to function but so too are muscles. There are a number of muscles that dorsiflex, plantarflex, evert, and invert the foot at the ankle. All of these muscle also act to stabilize the joints they cross during locomotion. The muscles acting at the ankle and foot attach as high on the leg as the femoral condyles (thus crossing two major joints and having both a proximal and distal function) and as low as the distal tibia and fibula. Those muscles that arise from sites outside of the foot and that attach at some point(s) within the foot will be considered extrinsic muscles and grouped into anterior, posterior and lateral groups. The muscles having both attachment sites in the foot will be considered intrinsic muscles. As they are quite numerous and their major function is simple movement of the phalanges, they are not presented. The three groups presented here are the extrinsic anterior, lateral, and posterior groups. Why no extrinsic medial group? Reach down and feel along the medial surface of your tibia. What do you feel? You should say bone. The medial surface of the tibia is quite superficial, that's why Muay Thai fighters and mixed martial artists use the tibia (shin) as a kicking weapon. It is an extremely hard surface that can be moved quite powerfully.

The anterior muscles acting at the ankle represent a small functional muscle mass that produce dorsiflexion of the foot at the ankle as they run in front of the talocrural joint. They provide some extension (lifting) of the toes as well. Dorsiflexion is important in posture maintenance, walking, running and any other movement requiring the forefoot to be elevated. Like so many other posture dominant muscles, these are predominantly slow oxidative and fast oxidative-glycolytic muscles. While there are exercises that can isolate these muscles in a weight training program, this is likely a rather poor utilization of training time. In fact, Bill Pearl a former Mister Universe and author of many respected books on body building did not prescribe any such exercises in workouts for bodybuilders, athletes, or the general public. The ankle musculature is automatically loaded and strengthened with any weight bearing exercise that requires balance (barbell squat for example). The more weight lifted, the more demand for the anterior musculature to generate force thus the stimulus for development is innate. In terms of endurance development, simple uphill running is quite a significant stressor, adequate for driving metabolic adaptations for endurance.

Tibialis Anterior – Nomenclature cues make localizing this muscle elementary, it is found along the anterior surface of the tibia, specifically the anterior and lateral surface. The muscle is quite superficial and attaches proximally along a line from just inferior to the tibial plateau and medial to the tibial tuberosity down approximately two thirds of the length of tibia. It attaches distally to the medial cuneiform and the proximal and inferior surface at the base of the first metatarsal. This muscle is easily palpated. With the foot on the ground, place your fingers on the lateral surface of the tibia, a few centimeters lateral to the tibial tuberosity and elevate the forefoot while keeping the heel on the ground (dorsiflexion). You will feel the muscle's apparent contraction. Repeat the contraction-relaxation cycle as you walk your fingers down the length of the muscle. As you approach the distal portion of the tibia, you will feel the muscle mass give way to the more rigid tendinous portion just superior to the ankle. You may be able to palpate the tendon to its distal attachment, but care must be taken as several other tendons are in very close anatomical proximity and their muscles are active simultaneously with the tibialis anterior. As the orientation of the muscle angles across the lower leg lateral to medial, you can surmise that the muscle will also assist in inversion at the ankle.

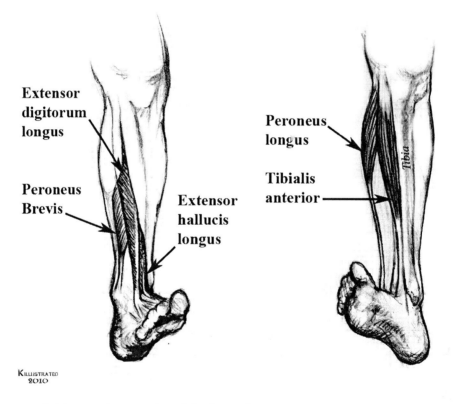

Figure 9-28. Anterior muscles of the lower leg.

Extensor Digitorum Longus – Although not as readily apparent as the nomenclature clues in the tibialis anterior, there are some helpful hints found in this muscle's name. The term extensor in the name implies the muscle and/or tendon will course to the anterior of the talocrural joint and digitorum suggests the distal attachment will be at the phalanges. The extensor digitorum longus lies posterior and lateral to the tibialis anterior. It attaches proximally along a narrow line from the lateral condyle of the tibia, down along the upper three fourths of its length, then attaches distally to the bases of the second and third phalanx of the second through fifth phalanges. This muscle is fairly easy to palpate. With the foot on the ground, place your fingers on the to the lateral and posterior to the tibialis anterior elevate the forefoot while keeping the heel on the ground (dorsiflexion). Try to discriminate between the active anterior musculature and the inactive posterior musculature, as the upper portion of the extensor digitorum longus is covered in part by the posterior gastrocnemius and peroneus longus. You may not be able to feel the muscle's apparent contraction until about one half the way down the tibia. As you approach the lateral malleolus, you will feel the muscle mass give way to the more rigid tendinous portion just anterior to the ankle. You may be able to palpate the tendon to its distal attachment, but care must be taken as several other tendons are in very close anatomical proximity and their muscles are active simultaneously with the tibialis anterior. The distribution of the distal tendinous attachments across the second through fifth phalanges should suggest to you that these are toe extensors. Their placement across the foot indicates little role in eversion or inversion.

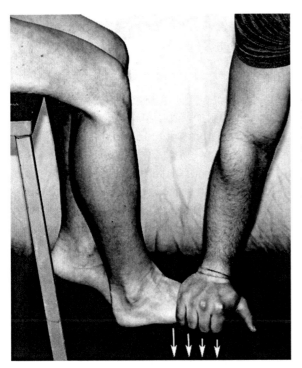

Figure 9-29. To get the anterior muscles of the to become more prominent, simply apply a downward force to the top of the foot, against a contraction. You can then palpate relative to bony features.

Extensor Hallucis Longus – There are two major clues in the name of this muscle. Hallucis refers to the hallux (big or great toe) and extensor indicates that the muscle will extend or lift the toe, thus the tendon will be anterior to the ankle and on the foot's dorsal surface. The muscles proximal attachment is the central half of the fibula and attach distally at the base of the second phalanx of the first phalange or hallux. The muscle itself usually cannot be palpated by beginners as both the tibialis anterior and the extensor digitorum longus lie over the majority of the muscle. With careful palpation of the anterior tendons during dorsiflexion, you can isolate the extensor hallucis longus tendon. Simply start by palpating the tibialis anterior tendon, then walk your fingers laterally and locate three side-by-side tendons. The tibialis anterior tendon is the most medial, the extensor hallucis longus tendon is slightly lateral, and the extensor digitorum longus tendon is the most lateral of the three. Owing to its mildly diagonal track from the fibula to the fist phalange, the flexor hallucis longus can contribute to inversion of the foot.

There are two relatively small extrinsic lateral muscles acting at the ankle, the peroneus longus and peroneus brevis. Both of these muscles act to plantar flex and evert the foot. These lateral muscles are primarily postural in function, assisting in maintaining equilibrium both anterio-posteriorly and medio-laterally. As such, they are predominantly slow oxidative and fast oxidative-glycolytic muscles. While there are exercises that can isolate these muscles in a weight training program, this is likely a rather poor utilization of training time. There are also running shoes marketed for individuals that evert excessively during their stride, BUT these shoes are designed around the heel-strike technique. Fix your technique and they aren't needed.

Peroneus Longus – It may be easier to remember the alternate name to this muscle in order to locate it, the fibularis longus. Its proximal attachment runs from just inferior to the fibular head and down approximately one half the fibula's length. The tendon is quite long and runs behind the lateral malleolus and attaches distally at the first metatarsal bone and on lateral side of the medial cuneiform (crosses the bottom of the foot at about a 45 degree angle – this is why it is an everter). This muscle is quite superficial and can usually be visually identified in its location between the tibialis anterior and the gastrocnemius (a posterior muscle). You can identify this muscle in two ways, either evert the foot, identify the muscle belly then walk your fingers down the length of the muscle and tendon around the ankle or start with the easily palpated tendon at the ankle and walk your fingers up to the muscle.

Peroneus Brevis – Like the peroneus longus, the brevis might be easier to locate using the alternate name, fibularis brevis. Frequently the pair of peroneus muscles are considered as a single unit as they are extremely close in anatomy and function. The brevis attaches proximally along the lower two thirds of the fibular shaft with the tendon becoming apparent just superior to the ankle. The tendon passes to the posterior and inferior aspects of the ankle and attaches distally at the base of the fifth metatarsal. The peroneus brevis has a primary function as a plantar flexor but does contribute some to eversion of the foot. The muscle is not palpable as it lies under the peroneus longus. The tendon from the ankle to the base of the fifth metatarsal may be palpable is some individuals during combined eversion and plantar flexion.

The extrinsic posterior muscles of the lower leg (the calf) are the largest and most powerful muscle mass of the three extrinsic groups. Their primary function is plantar flexion with each muscle having additional movement roles. These muscles are heavily oxidative with 50% or more of their fibers being distinctly slow oxidative and a majority of the remaining half being fast oxidative glycolytic. Their fiber type profile is indicative of a strong postural function but their mass relative to the other two extrinsic groups indicate an equally strong ambulatory function. The previous discussion of footwear and function supports this contention. An interesting observation arises here regarding the comparative structure and function of the three extrinsic muscle groups and running technique. The posterior musculature is adapted extremely well to generating concentric propulsive force and extremely well adapted to eccentric force absorption. Including six muscles of the posterior muscle group, there are eight plantar flexors of the ankle and foot and only four dorsiflexors. So given the unique supportive arch structure of the foot and the abundance of musculature to produce toe off propulsion and forefoot shock absorption, and given the lack of calcaneal shock absorptive capacity along with a lack of muscle mass dedicated to dorsiflexion, it seems logical that the body is structured to run with a forefoot strike rather than a heel strike technique.

Plantar flexors can be seen as part of virtually any isolation type weight training program for good or ill. Calf raises, donkey raises, and seated calf raises are all common exercises (figure 9-30). These exercises can be effective in increasing mass and strength, but if we consider what occurs in weighted movements like the squat, deadlift, or even a power clean we will see plantar flexion strongly represented as part of the exercise (figure 9-31). Again, training efficiency is an important consideration in programming. If a multi-joint exercise works the muscle group effectively at the same time it works others, do we really need to isolate a muscle or a muscle group?

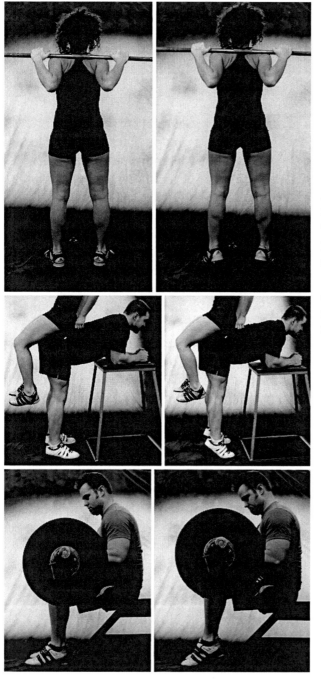

Figure 9-30. Calf raise, donkey raise, and seated calf raise (top to bottom). The calf raise is the most useful and comprehensive calf exercise (referring only to isolation exercises). The donkey raise is the silliest and most limited. The seated calf raise is the most uncomfortable.

Figure 9-31. Ankle movement in the squat. Note the angle changes of the ankle joint, requiring of a strong contribution, concentrically and eccentrically, of the posterior ankle musculature.

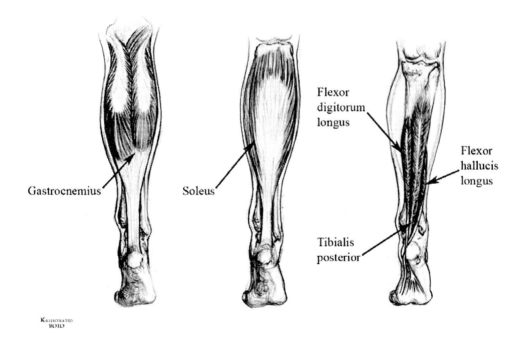

Figure 9-32. Posterior muscles of the lower leg, superficial to deep (left to right).

Gastrocnemius – The gastrocnemius, frequently referred to as a gastroc, is the largest muscle of the posterior group. It is pennate and has two distinct heads and a distal tendon of attachment it shares with its underlying musculature. Its proximal attachments occur with each muscular head, medial and lateral, attaching at the posterior surface of the medial and lateral femoral epicondyles, just superior to the condyles. Both heads fuse into a single distal tendon that attaches to the posterior and anterior surface of the calcaneous. This distal tendon is commonly called the Achilles tendon but can also be called the calcaneal tendon. The gastrocnemius crosses two joints and as such has both a proximal and distal function. The acknowledged proximal function is flexion of the knee. If the foot is the least stable end of the system this is correct and the proximal action will be lower leg movement rearwards in an arc around the knee. If the foot is immobile, as it is during most of our ambulatory life, the proximal function will be to assist in "locking out" the knee, pulling the distal femur into vertical alignment with the proximal tibia. The accepted distal function of the gastrocnemius is plantar flexion at the ankle with the foot pushing a resistance down or pushing the body up. This is occurs with every stride of locomotion we take and is part of many exercise and sport movements. However if the foot is held immobile, flat on the floor, then the distal function is strictly postural and pulls the mass of the body to the rear – to cooperatively balance the actions of the anterior musculature to maintain the center of gravity over the base provided by the feet. If the knee is bent, the gastrocnemius is limited in its ability to produce force. Since knee flexion reduces the distance between proximal and distal attachments, the sarcomeres present cannot shorten sufficiently to produce high force (this is a physiological concept – the length tension curve). You can test this out on someone, have them stand up and go on their toes as hard as they can then have them sit down and go up on their toes again. Press your fingers into the bellies of the gastrocnemius in both conditions and you will notice a difference in muscle rigidity.

Soleus – The soleus lies beneath the gastrocnemius and is a fairly thin muscle but it is almost as wide as the gastrocnemius. In lean individuals it can be seen protruding slightly just inferior and anterior to the major bellies of the gastrocnemius. The soleus attaches proximally to both the tibia and the fibula. The attachement sites form an inverted "V" on the upper quarter of the tibia and the upper third of the fibula. Distally the tendon of the soleus fuses with that of the gastrocnemius to form the calcaneal tendon. The soleus crosses one joint, the ankle, and by virtue of its central orientation, if the foot is the least stable structure, it will produce plantar flexion as with the distal function of the gastrocnemius. If the foot is held fixed to the floor it will act as a postural muscle and pull the mass of the body posteriorly. This latter role is the primary function of the soleus as it contributes greatly to maintaining standing balance. It is comprise of 60% or more of slow oxidative fibers making it well adapted to its function. Palpating this muscle depends on your ability to palpate the separation between the inferior border of the gastrocnemius and underlying soleus. You can use the knee flexion trick from the gastrocnemius above to help. Bend the knee to 90 degrees and elevate the heel firmly. The soleus will be the prime mover and more rigid than the gastrocnemius. It is relevant to note that you will see references in the literature to the "triceps surae." This term refers to a more antiquated grouping of the gastrocnemius (two heads) and soleus into one muscle group.

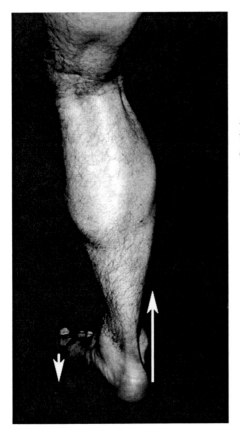

Figure 9-33. To make the posterior muscles of the lower leg, simply raise the heels and go up on the toes.

Flexor Digitorum Longus – The flexor digitorum longus muscle is, as the name implies, a long one. It is also a flexor of the digits, meaning that it will pass behind the medial malleolus and attach inferiorly to the phalanges. The muscle attaches proximally on the middle third of the medial aspect of the tibia, its central tendon will pass behind the medial malleolus then angle laterally and split into four slips of tendon attaching to the base of each distal phalanx of phalanges two through five. The muscle is deep and cannot be palpated but the tendon of the flexor digitorum longus can be palpated just inferior to the medial malleolus during repeated strong flexion of the toes. It is also possible to palpate the tendon distal to the metatarsophalangeal joints on the plantar surface of the lateral four digits. Although it is named according to its function to flex the toes, the primary function of the flexor digitorum longus is to plantar flex the ankle during locomotion. As with other members of the extrinsic posterior group, it is also active in postural control when the foot is the most stable structure.

Tibialis Posterior – The nomenclature yields the clues needed to localize this muscle located on the posterior surface of the tibia. The name is not completely descriptive as the proximal attachment of the tibialis posterior is divided between the lateral aspect of the shaft of the tibia, the medial aspect of the fibula, and the interosseus membrane from the level just inferior to the head of the fibula to approximately half way down the length of the fibula. After the tendon passes behind the medial malleolus, the distal attachment is quite complex with connections on the plantar surface of seven bones; the navicular, all three cuneiforms, and the second, third, and fourth metatarsals. The muscle is quite deep but you can identify the tendon by strongly plantar flexing and inverting the ankle and subtalar joint simultaneously, then palpating the area just posterior to the medial malleolus. Aside from the plantar flexion and inversion functions, the tibialis posterior is quite involved in both anterio-posterior and medio-lateral postural control.

Flexor Hallucis Longus – Yet another muscle with great clues in its name. As you would expect, the flexor hallucis longus flexes the hallux. "Flexor" indicates it is a posterior group muscle. The muscle attaches proximally on the posterior of roughly the middle third of the fibula it then crosses the tibia and passes behind the medial malleolus. It attaches distally at the base of the distal phalanx of the hallux. As it passes behind the medial malleolus, it contributes to plantar flexion as a function along with its namesake function of flexing the hallux. Palpation of the muscle is not possible, but the tendon can be palpated on the plantar surface of the foot. Place your fingers along the length of the first metatarsal then repeatedly flex and extend the big toe (it doesn't matter if all your toes move). You will note that the tendon is most notable when the tendon is stretched during extension (toes up).

10 – THE KNEE

Walking, jumping, kicking, running, lifting – pretty much every task we take for granted as a normal part of human existence relies, at least in part, on the knee. The knee joint is a slight modification of a simple condyloid hinge joint that allows extension and flexion – the modification is that it also allows for a small amount of lateral and medial rotation. The bones of articulation in the knee are the femur of the thigh, the tibia and fibula of the lower leg, and the patella (or kneecap). The condyles of the femur sit atop the tibial articular surfaces. The patella is located just anterior and superior to the femoral condyles. The fibula articulates with the medial tibia and the lateral aspect of the femur above it. It is an easy bony architecture to learn and remember.

Imagine a leg without a knee and the Frankensteinish walk that would result. Imagine trying to lift something from the floor. Imagine sitting down or worse, standing up. The knee is a very important joint and one of the most frequently injured joints in the human body – through acute injury or through chronic overuse without proper pre-conditioning. Ranging in severity from simple soreness to architectural destruction, knee injury affects the young and old, male and female, the athletic and the sedentary. About 33% of the population 45 and older report knee pain of some type. Two percent of the US population over age 55 has pain severe enough and arthritis severe enough to consider arthroplasty (knee replacement). Athletes have a high incidence of knee problems but they are not the only fit group with problems, nor do they have the highest frequency of knee problems. Ninety five percent of ballet dancers will report knee pain at some point of their career.

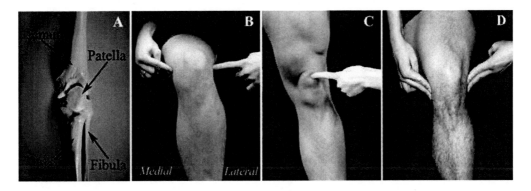

Figure 10-1. The bones of the knee (A) can be easily palpated. Superior and lateral to the bend in the knee, the femoral epicondyles can be felt (B). The patella can be seen and felt on the anterior aspect of the knee (C). The medial tibia and the lateral fibula can be palpated below the bend of the knee.

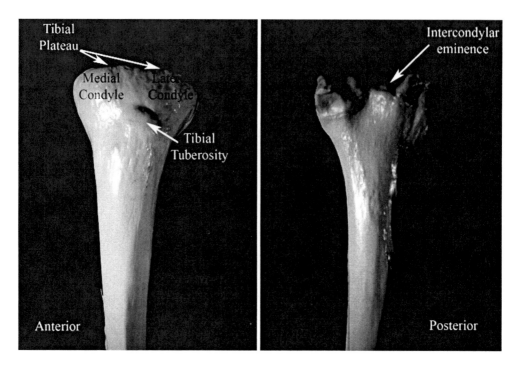

Figure 10-2. Bony features of the proximal tibia.

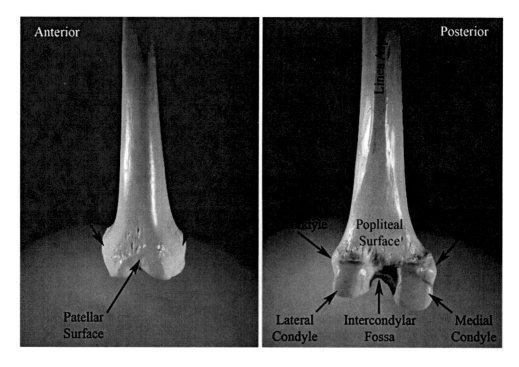

Figure 10-3. Bony features of the distal femur.

CONNECTING THE BONES

There are a number of connective tissue structures in the knee – ligaments, tendons, menisci, and bursa. Given the limited stability and support provided by the skeletal elements and the tremendous amount of joint use over a life span, these structures are quite important.

Patellar Ligament – This ligament is sometimes a confusing structure. But it shouldn't be. The patellar ligament connects the inferior surface of the patella to the tibial tuberosity (figure 10-4). Bone connected to bone. Occasionally some sources will claim that the patella is a sesamoid bone (bone imbedded in a ligament) and refer only to a patellar tendon connecting the combined tendons of the quadriceps to the tibial tuberosity. Careful examination of the structure and centuries of study by many many amazing anatomists support the presence of the patellar ligament (and a separate patellar tendon described later). Palpating the tendon is quite easy. With the leg relaxed and the knee at ninety degrees, place your finger at the inferior and middle aspect of the patella then using moderate pressure walk your fingers down to the tibial tuberosity. You should feel a slightly flat and rebounding structure about an inch wide between the points of attachment.

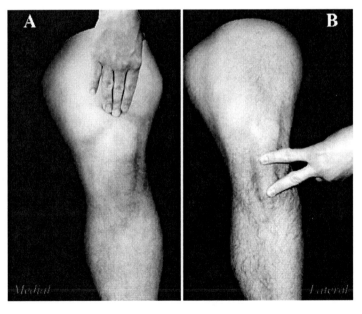

Figure 10-4. The patellar tendon (A) and patellar ligament (B) can be palpated immediately above and below the bone.

When the knee is flexed, the patellar ligament lies in the groove between the two femoral condyles. If the groove or the patella are anatomically abnormal, a patellar tracking problem (the patella moving outside of the groove) may occur and cause pain. Despite the attention this condition receives from some professional groups and the plethora of patellar tracking correction devices available, it is in truth a relative rarity and is not present in normal anatomy during exercise. It is usually associated with at least a partial tear of one of the ligaments enveloping the patella. Most frequently the tibial collateral ligament (aka the medial patellofemoral ligament) is injured - in more than ninety percent of the diagnosed cases. A patellar tracking injury may occur if an impact dislocates the patella from its normal position (figure 10-5a). Contrary to many contemporary myths foot angle during exercise does not induce patellar tracking injury, nor will simple valgus (knees in) or varus (knees out) flexion and extension of the loaded knee. These conditions may cause knee irritation and pain over time, but will not force the patella out of its groove. This statement is also valid for complete range of motion squats. During the first few degrees of knee flexion there is a mild instability and the patella will rotate on the surface of the femur, this is normal and allows the patella to seat itself. As flexion continues, the patella slides into the deepest part of the crevasse between the two condyles and becomes very stable. The simple contraction of a loaded quadricep acts to pull the ligament firmly into the groove, reinforcing the anatomical stability of the patella throughout the complete knee flexion/extension range of motion (figure 10-5b). A simple safeguard for knee health is to ensure that knee flexion and extension occurs with the femur and tibia tracking in a line following the long axis of the foot. This minimizes rotation of the femur on the top of the tibia and keeps the forces applied to the patella symmetrical.

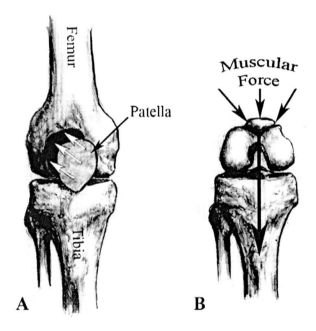

Figure 10-5. Patellar tracking injury moves the patalla out of the channel between the femoral condyles (A). Contraction of the knee musculature pulls the patella into the channel and minimizes risk of patellar displacement (B).

Collateral Ligaments – There are two collaterals ligaments, a medial and a lateral that provide lateral stability to the knee by reinforcing normal muscular tone in preventing excessive side-to-side movements. The two ligaments are also known as the tibial and fibular collateral ligaments. The medial collateral ligament connects the medial epicondyle of the femur and the medial tibial surface. The lateral collateral ligament runs between the lateral epicondyle of the femur and the head of the fibula. You can find both of these ligaments by placing your fingers on the sides of the knee and repeatedly flexing and extending the knee. Once you discriminate the gap between the femur and the tibia and fibula, with a little pressure you should be able to feel a fairly resilient and rebounding cord. The collaterals are injured frequently by lateral impacts to the opposite side of the ligaments location (eg., a blow from the medial side stretches and disrupts the structure of the lateral collateral).

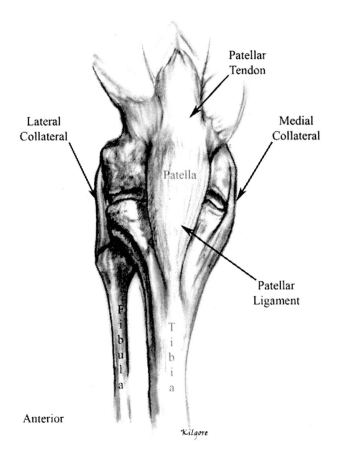

Figure 10-6. The patellar and collateral ligaments. Please remember that ligaments impart stability during movement only if the musculature fails to do so.

Cruciate Ligaments – Again there are a pair of ligaments operating here that act as limiters of joint separation, the cruciates. The anterior cruciate ligament attaches at the lateral condyle of the femur and to the anterior intercondylar surface of the tibia. It acts to limit the excursion of the tibia too far forward (or femur too far backward) if muscular tension is inadequate. The posterior cruciate runs from the medial condyle of the femur to the posterior intercondylar surface. It prevents displacement of the tibia to the posterior (or femur too far forward). The cruciates are also responsible for limiting the amount of rotation possible along the long axis of the knee. In plant-and-twist sports, such as soccer, football, basketball, volleyball, rugby, ad nasueum, there is an elevated risk of injuring these ligaments as compared to linear sports such as running, cycling, lifting, etc. Although there are approximately 80,000 anterior cruciate ligament injuries in the US each year, only about 30% of them are the result of forceful contact. Of those injuries related to sport and exercise, the majority occurs between the ages of 15 to 25. Approximately 1 in 100 high school and 1 in 10 college female women will injure the anterior cruciate each year. It is estimated that the medical revenues off of these injuries from this population alone, tops $100,000,000 per year. Such injuries are not career ending as surgical repair techniques are quite effective and robust. Surgery can be avoided by effective prevention as the risk of anterior cruciate injury can be reduced by up to 88% with effective strengthening of the complete musculature around the knee – and yes, that means lifting weights is good for your knees. Although this information has been known for more than three decades, the value and advice to strengthen the knee musculature is frequently ignored by coaches and clinicians or worse, baselessly contradicted.

Menisci – The term meniscus is a quite broad term derived from the Greek and means a crescent shaped curved surface (convex or concave), usually used in reference to the curved interface between a liquid in a tube and the atmosphere. However, in the knee a meniscus is one of two roughly horseshoe shaped cartilages located on the articular surfaces of the tibia, with the horseshoes pointed towards the middle of the joint. The menisci are thicker towards the outer borders and thinner as they approach the middle. The shallow bowl like shapes formed via their position atop the tibia help to provide a modicum of stability to the knee, as the condyles of the femur sit in the bowls. Their primary function however is to reduce friction between the femur and tibia during movement. The menisci are intimate with the cruciate ligaments. The lateral meniscus in fact has a ligament, the meniscofemoral ligament that parallels the larger posterior cruciate. The menisci are often torn at the same time cruciate or collateral ligaments are injured.

Bursae – A bursa is a fluid filled sac bounded by fibrocartilage and situated around various condyloid joints like the knee, usually just underneath tendons and ligments. They vary in size and function to cushion the joint around which they are situated. An example would be the prepatellar bursa which lies just behind the patella and helps maintain the position of the patella and dampens forces between the patella and the underlying femur.

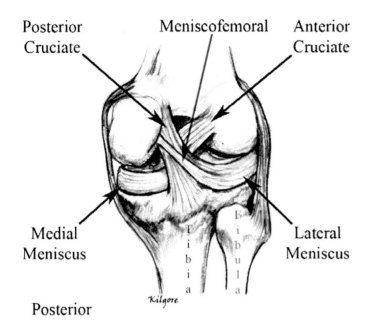

Figure 10-7. The cruciate ligaments, meniscofemoral ligament, and the menisci.

Muscles

Muscles acting at the knee are generally categorized into two groups: anterior extensors and posterior flexors. Grouping the muscles like this makes learning each muscle's function simple.

The anterior muscles, the vastus medialis, vastus lateralis, vastus intermedius and rectus femoris are commonly known as the quadriceps. The vasti (plural) attach along the length of the femur and run to the tibial tuberosity via the patellar tendon, the patella, and then the patellar ligament. As mentioned earlier, there seems to be some debate among exercise professionals if there is actually a patellar ligament or not BUT the *Anatomica Nominica* and the International Congress of Anatomists are the ultimate authorities and say there is. These muscles are quite strong and their orientation to the joint, a very short lever arm wrapping around the axis of rotation, enables large-scale movement of the lower leg with a small amount of muscle shortening. This is an excellent combination, large muscle masses producing force and an advantageous lever arrangement for velocity, making for powerful movement.

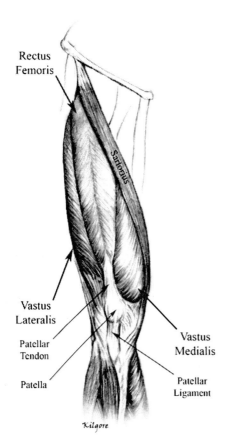
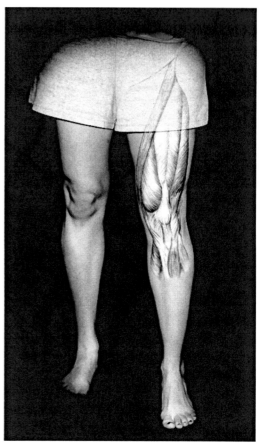

Figure 10-8. The anterior musculature of the knee, the rectus femoris, vastus medialis, and vastus lateralis. The vastus intermedius lies under the rectus femoris.

Rectus femoris – The rectus femoris covers a large portion of the middle of the thigh. It completely covers the deep vastus intermedius and a large portion of the other two vasti. In a lean individual a strong knee extension will reveal an apparent V-shaped muscle belly lying between and generally superior to the vastus medialis and vastus lateralis. It has two proximal attachments, the anterior inferior iliac spine and just superior to the acetabulum (where the femur articulates at the hip). Use of the term rectus in its name implies a vertical and straight orientation along the femur. The rectus femoris attaches distally to the tibial tuberosity by way of the patellar tendon-patella-patellar ligament series. The rectus shares its distal function with the vasti, knee extension. Its contribution to knee extension is strongest when the hip is extended (straight) as a flexed (bent) hip invokes the muscles proximal function, hip flexion. With a flexed hip, the rectus is already pre-shortened and cannot greatly contribute to knee extension. This is one reason that an isolation exercise, like the machine leg extension, is not adequate to fully develop strength in the knee musculature.

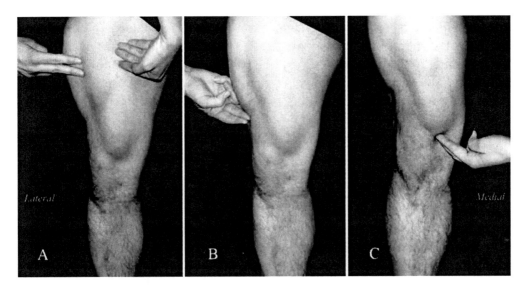

Figure 10-9. Palpation of the anterior and superficial knee muscles, the rectus femoris (A), the vastus lateralis (B), and the vastus medialis (C), is fairly simple.

Vastus lateralis – This muscle is sometimes called the vastus externus and is proximally attached to the femur along to the anterior and inferior borders of the greater trochanter, the lateral aspect of the gluteal tuberosity, and to the upper portion of the lateral side of the linea aspera. The linea aspera is a long vertical ridge on the posterior surface of the femur. The muscle also attaches distally to the lateral portion of the patella and then to the tibial tuberosity. It has but one function, to extend the knee. A simple strong extension of the knee will make the vastus lateralis visually apparent as a prominent muscle belly about two inches superior and lateral to the knee.

Vastus medialis – This muscle is attached proximally from the inner side of the neck of the femur, along the entire length of the linea aspera, and to the medial, anterior, and lateral aspects of the upper three fourths of the femur. It attaches distally to the medial portion of the patella and then to the tibial tuberosity. It has an identical function to the vastus lateralis, extension of the knee. A simple strong extension of the knee will make the vastus lateralis visually apparent as a prominent tear-drop shaped muscle belly at the level of the superior border and medial to the patella. You will see this muscle referred to as the VMO frequently in popular fitness and some clinical literature - it's short for vastus medialis obliquus, obliquus referring to the angle of the muscle fibers in the muscle.

Vastus intermedius – Sometimes referred to as the vastus cruraeus (derived from Latin and referring to the lower leg) this muscle attaches proximally along the anterior and lateral surfaces of the upper two thirds of the femur. The intermedus lies under the superficial rectus femoris. The muscle attaches to the superior under-portion of the patella by way of the patellar tendon then to the tibial tuberosity. The vastus intermedius is a deep muscle and is not palpable but it is active in the same action as the other two vasti, knee extension.

If you wish to verify what these anterior muscles do, you can use an involuntary reflex and force the muscles to demonstrate their function. Have someone sit down with his upper leg supported to the back of the knee and the lower leg hanging free. Take a reflex hammer or used the side of your hand (karate chop style) and give him a firm tap on the patellar ligament. The reflex invoked by activating the stretch receptors will extend the knee from a few to several inches – the basic function of the anterior knee musculature. After you are successful in producing the reflex action, you can explore the involuntary nature of the reflex. Have your subject try to stop the reflex by flexing and holding knee with a strong hamstrings contraction. Most likely, he will not be able to stop the extension. In fact, the reflex may become exaggerated in some individuals.

The posterior muscles are more numerous and more of them cross two joints than seen in the anterior musculature. Many will also be considered in the hip chapter. There are several large muscles in this group and several smaller muscles.

Biceps femoris – This is the muscle most people will equate to the "hamstrings" on the back of the leg, however there are two other muscles, the semimembranosis and the semitendinosis, that comprise the hamstrings. The biceps femoris, as its name indicates, has two heads. The long head has a proximal attachment on the ischial tuberosity (seat bone) and on the lower part of the sacrum. The short head attaches along the upper posterior aspect of the linea aspera of the femur. Afterwards the muscle runs down, medial to lateral, across the back of the leg. Its distal tendon attaches to the head of the fibula and to a lesser degree the lateral condyle of the tibia. The tendon of the biceps femoris is closely associated with the lateral collateral ligament of the knee. The short head of the muscle is a knee flexor. The long head crosses both the knee and hip joint so its distal function is knee flexion and its proximal function is hip extension. If the knee is flexed, the biceps femoris is pre-shortened and is a weak hip extensor. If the knee is extended, the biceps femoris is a weak hip flexor. In human movement, there are very few absolutes and there will be a significant amount of variability in the degree of muscle involvement in the various tasks utilizing this muscle. You can palpate the tendinous attachment to the head of the fibula by flexing the knee to ninety degrees against resistance. Place your fingers on the head of the fibula and walk your fingers posteriorly along the very pronounced tendon until it dissipates into the muscle.

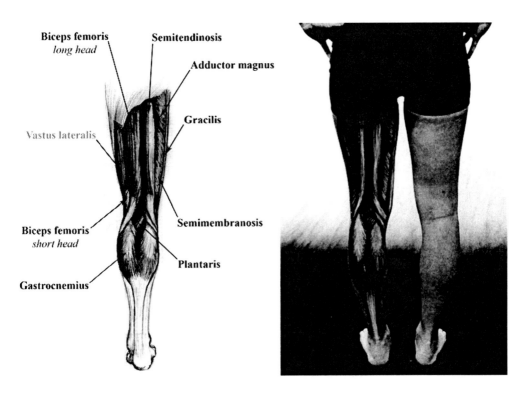

Figure 10-10. The posterior musculature of the knee.

Semimembranosis – The second component of the hamstrings, the semimembranosis, attaches proximally at the ischial tuberosity and attaches distally to the posterior aspect of the medial condyle of the tibia. It runs medial to the biceps femoris. The muscle crosses two joints and acts at both the hip and knee. The distal function at the knee is flexion of the knee. You can palpate the tendinous attachment to the medial tibial condyle by flexing the knee to ninety degrees against resistance. Place your fingers on the medial aspect of the tibia until you find two prominent tendons. Follow the most inferior of the two and follow it medially and upwards until you find the bulk of the muscle.

Semitendinosis – The final element of the hamstrings is the semitendinosis. The muscle attaches proximally at the ischial tuberosity and shares much of its attachment architecture with the biceps femoris. The semitendinosis is notable in that the muscle ends at approximately mid-thigh and attaches distally to the medial upper face of the tibia by way of a long tendon. Many of the flexors of the knee have tendons that attach in a type of wrap-around orientation to the tibia. The semitendinosis runs around the medial condyle and insinuates itself over the medial collateral ligament before its final termination. The semitendinosus has a proximal hip function but its distal function at the knee is flexion. You can palpate the tendinous attachment to the behind the medial tibial condyle by flexing the knee to ninety degrees against resistance. Place your fingers on the medial aspect of the tibia until you find two prominent tendons. Follow the most superior of the two and follow it medially and upwards until you find the bulk of the muscle. You will note also that tendon is also slightly more lateral than that of the semimembranosis.

Sartorius – The sartorius is a very long and thin muscle that crosses diagonally, lateral to medial, the anterior length of the thigh – it is the longest muscle in the human body. Its body crosses over the upper portion of the rectus femoris, runs over the middle portion of the vastus medialis and then borders it medially until their attachments distally. The muscle proximally attaches at the anterior superior iliac spine, passes behind the medial condyle of the femur and attaches distally to the proximal, medial, and posterior surface the tibia. The sartorius has four movement functions with three of those occurring at the hip. At the knee however it is simple flexor.

Popleitis – This is a deep and very small muscle that attaches proximally at the lateral femoral condyle and attaches distally to the posterior tibia below the medial tibial condyle. It crosses the back of the knee lateral to medial. Given its small mass and attachment orientation it is a weak knee flexor. The poplietis also acts on the lateral meniscus by way of a section of its tendon attaching to the posterior lateral meniscus. When the knee is flexed actively the popleitis draws the meniscus to the posterior and prevents pinching of the meniscus between the tibia and femur. There is an surface landmark of interest here, the poplietal fossa, the depression on the back of the knee. Its superior border is formed by the transition from hamstring muscles to tendon, to the inferior by the vergence of the gastrocnemius, and by the tendons attached to the lateral and medial tibial condyles. You can palpate the fossa but the muscle is deep and cannot be palpated.

Plantaris – The plantaris is a relatively famous muscle in the muscle hypertrophy (increased muscle mass) research community. It is quite large in some animals, such as the great apes, but it has little function in modern human beings and may be completely absent in up to ten percent of the population. The plantaris in the mouse and rat are frequently used in exercise physiology experiments that explore the mechanisms of muscle hypertrophy through the ablation of the gastrocnemius. When the superficial gastrocnemius is removed, the underlying plantaris and soleus must pick up the load once carried by the gastrocnemius. As a result they hypertrophy. In the mouse there also relatively few cells in these muscles (in the thousands) so it easier to enumerate and measure the cells. You can see by the close proximity of the two attachment sites, that the muscle and tendon together will not be more than about four inches long. The primary, and proximal, function of the plantaris is knee flexion but its tendon also connects to the calcaneal tendon of the gastrocnemius. It thus has a distal function as a weak plantarflexor.

Gastrocnemius – The gastrocnemius' proximal function, flexion of the knee, is of interest here. Its lateral and medial heads attach to proximally and respectively to the lateral and medial condyles of the femur. It attaches distally to the calcaneus via the calcaneal or Achilles tendon. Its distal function, plantarflexion of the ankle, and its palpation were discussed in the previous chapter.

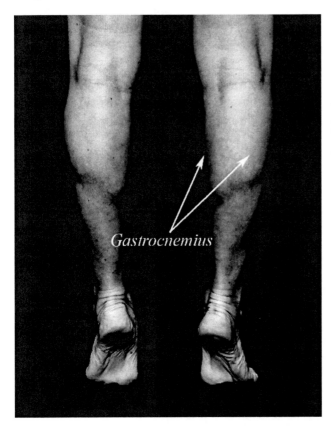

Figure 10-11. The gastrocnemius is easily visualized on the posterior aspect of the calf.

An Agonizing Issue

As discussed in the *Force the Issue* chapter, the relationship between agonist and antagonist muscles and muscle groups is an issue of quite a bit of interest. This interest is in fact not new. In 1907, W.P. Lombard described an anatomical and physiological paradox (a statement that contradicts itself), in that co-activation of the quadriceps and hamstrings – an antagonistic set of muscles – worked cooperatively to move the body from sitting to standing. It was baffling at the time that opposing muscle groups acted together to produce movement since simultaneous antagonistic muscle group activity would seem, logically, to produce an isometric muscle action (no movement). How could the opposing knee extensors and knee flexors be active at the same time to produce movement?

Lombard and many subsequent researchers evaluated the paradox within the context of isolated movement, muscles, and joints and none have provided any substantive explanation of how the body copes with the paradox. But it does. Within this text it is proposed that systems and most stable structure approaches are the most relevant towards explaining the how of human movement. The body is immobile at the foot during the act of standing and thus the kinetic chain starts there and continues through the vertebral column. It does not exist as an entity comprised of only the hip and knee, as it is frequently described. This paradox of the quadriceps and hamstrings, first described over 100 years ago and still debated today, demonstrates that while our understanding of human anatomy and function is large and ever-growing there remains a great deal more to learn.

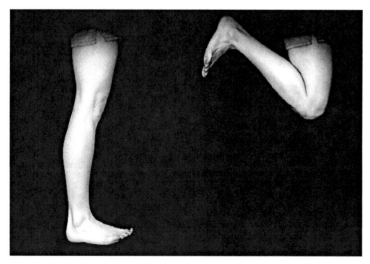

Figure 10-12. The knee in extension (left) and flexion (right). Anterior muscles generally extend and posterior muscles flex the knee joint, which end is most stable (the distal end above) affects the nature of what part of the body actually moves.

11 - THE HIP

The musculature around the hip represents some of the largest and most powerful muscles in the human body, the seat of power if you will. The extension of the hip can move thousands of pounds, it can propel the body up and over obstacles equal to or greater in height than the body in motion, and it can move the body at speeds greater than 25 miles per hour. But this is just the surface of the hip's unique abilities, it's bony and ligamentous structure provides for our standing posture at a low metabolic cost, its radical range of motion allows for mobility in virtually any anatomical plane or axis (figure 11-1), and its composite muscles are also capable of providing for ambulation of long distances and long durations. It is likely the most important target area for trainees and fitness professionals. Look at the list of activities listed earlier. Every one of those activities, each with the hips playing an integral role, is an imperative in daily life and in sport.

Figure 11-1. The hip allows a huge breadth of range of motion; flexion, extension, abduction, adduction, rotation, and circumduction.

BONES

The bony structure of the pelvis creates the foundation for the hip joint. The pelvis is formed by the fusion of the lateral ilium, lateral ischium, and the anterior pubis. The pelvis is completed to the posterior with articulations to the sacrum and coccyx. When considered as an intact complex, the pelvis is seen as an inward sloping structure (figure 11-2). The Latin meaning of "pelvis" is a basin.

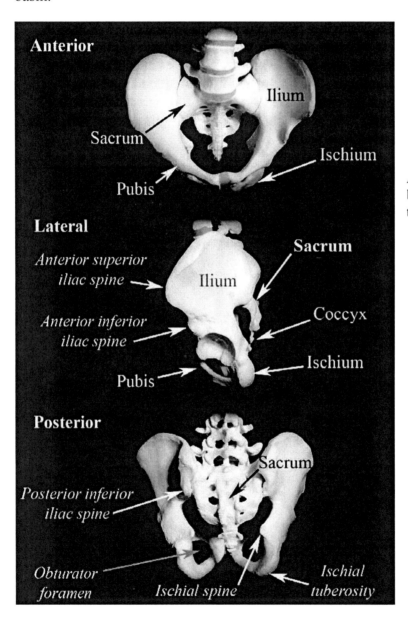

Figure 11-2. The bony structure of the pelvis.

The hip is a synovial, multi-axial, ball and socket joint, the largest of this type with a diameter in adults of approximately two inches. Formed by the articulation of the head of the femur and a fossa in the pelvis, known as the acetabulum (figure 11-3), the hip joint is a very mobile but fairly stable structure. Hip dislocations in sport are fairly rare.

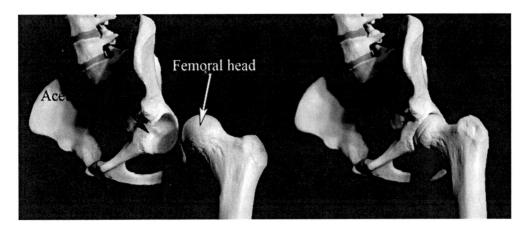

Figure 11-3. The hip joint is formed by the articulation of the acetabulum and the head of the femur.

The single most important aspect of the hip joint's structure is that it articulates with the proximal femur – a transition and connector between the vertebral column and the legs. The architecture of this joint enables the upper body to be supported while sitting, standing, and during ambulation. The pelvis provides many robust attachment points for muscles and ligaments. The pelvis structure also acts to protect soft tissues present (nerves, blood vessels, urinary tract, reproductive organs, etc.).

GRANDMA'S BONES

The femoral head attaches to the shaft of the femur via a narrowed neck that is prone to fracture in the elderly and women, due primarily to degenerative effects of a preventable pathology, osteoporosis. Osteoporosis, the loss of bone mineral content, leads to a weakening of the microarchitecture of bone making it susceptible to fracture from external forces. Think of it as osteo = bone and porosis = more porous or holey. Normal bone is much like limestone, extremely osteoporotic bone is much like gym chalk. The first stage of this pathology is osteopenia, the loss of active bone cells. This loss of cells that can produce new bone and repair damaged bone correlates well with the cessation of weight bearing activity. Osteoporosis affects approximately 24 million Americans (about 7.5% of the total population) with about 18 million of the affected being women over 45 years of age (about 12% of the total female population). If you examine activity patterns of American females, activity levels decay as age progresses, setting the stage for osteopenia. The adage "if you don't use it, you lose it" may be germane here. Consideration of bone loss during space flight and in paraplegia adds support to the accuracy of this ages old saying.

Osteoporosis imposes a significant medical expenditure load on private and publicly funded health care systems with an annual cost of direct medical treatment on the order of $15 billion. Beyond the financial burden there are many other costs for individuals; reduced quality of life, depression, loss of mobility, loss of independence, and mortality. In geriatrics, a fall and fracture of an osteoporotic hip or femur often means they will not leave the hospital or care facility.

In this book we are interested in the exercising and sporting populations, does osteoporosis affect them? It can. Osteoporosis is a fairly common pathology reported among female aerobics instructors. The excessively high volume of repetitive impact exercise, the limited choice and restricted calorie diets they follow, and limited recovery time they allow themselves sets the stage for the "female athletic triad" – amenorrhea (loss of menstrual cycle), eating disorder, and osteoporosis. This is a difficult situation, as instructors make their money instructing … teaching one or two classes a day won't pay rent in today's coaching and instructing pay scales. They have to overtrain in order to make a decent payday. There are no safeguards or surveillance methods in place for this group and unfortunately by the time there is a physical problem manifested, the bone damage is essentially irreversible.

There are two other populations that run the risk of early development of osteoporosis (and the female athletic triad), gymnasts and distance runners. The advantage here is that the national governing boards for these sports do have policies and educational programs in place in an attempt to reduce the frequency of occurrence – it is not in a coaches or sport organizations best interest to allow their talent to be permanently removed from competition.

Clinical communities spend much money and effort looking at identifying risk factors and testing methods to identify individuals who may develop osteoporosis in the future. NASA also spends great deals of money on developing interventions to bone loss during spaceflight.

You'll see lots of commercials on TV about taking this or that supplement to prevent osteoporosis. Yes, those supplements help a LITTLE but they can't cure or prevent the pathology because supplements don't induce a structural adaptation, they only provide a mineral resource. Bone growth is stimulated by an external load causing a mild deformation in the bone mineral architecture. A bent crystal, from having a load on you, produces the piezoelectric effect, or in other words causes an electrical charge to form (quartz crystal watches run on the piezoelectric effect). That charge stimulates bone cells to elaborate more bone materials to reinforce the bone's architecture. A very simple concept. So the very simple solution to the puzzle of osteoporosis is something clinicians and most women do not want to hear, get under a bar (the weight kind not the alcohol kind). The excessive catabolic cost of long distance running, the unloading effect of swimming, and every other let's-make-exercise-easy gimmick cannot produce an adaptive environment allowing for maintenance and reinforcement of bone.

In regards to bone and avoiding osteoporosis; don't beat it, don't float it, LOAD IT.

JOINTS

There is more than one joint in the hip region (the pelvic girdle). The hip bones (one set on each side) articulate to the anterior at the pubic symphysis (figure 11-4). This is a fibrocartilagenous joint normally capable of very very limited movement. During pregnancy, the hormonal milieu of the body changes and softens the cartilage and allows the pubic bones to move thus enlarging the birth canal. Disruption of the integrity of this joint under normally abnormal circumstances would be quite painful and debilitating, viciously landing pubis first on a saddlehorn or a top tube on a bike for example.

To the posterior, the ilium of each hip articulates with the sacrum of its respective side forming the much referred to sacroiliac joints (figure 11-5). These two articulations are very snug and movement here is unlikely. Irritation of this joint through trauma, repeated impact over very long durations (such as in running), or inappropriate and unbalanced loading can lead to inflammation and pain.

The hip joint proper, is a synovial joint formed by the articulation of the head of the proximal femur and the acetabulum (figure 11-3). The acetabulum is a bowl-shaped structure formed at the junction of the three pelvic bones — the ilium, pubis, and ischium. A little less than half the femoral ball sits in the acetabular bowl at any one time – the position of the femur dictates which half.

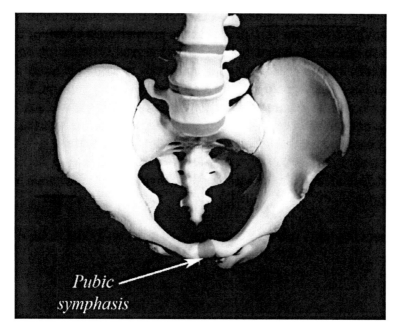

Figure 11-4. The pubic symphasis.

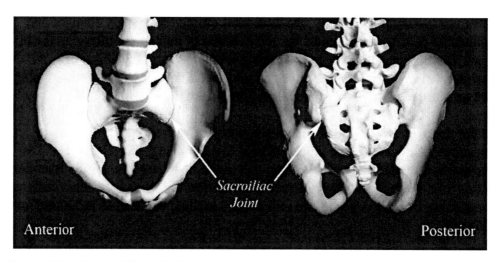

Figure 11-5. A sacroiliac joint is the articulation of the sacrum and the ilium.

There are five ligaments active at the hip. Four of these are extracapsular ligaments and are present on the outside of the acetabulum. The iliofemoral, the ischiofemoral, and the pubofemoral ligaments attach, as expected by virtue of the nomenclature, to the ilium, ischium, and pubis and then to the femur (figure 11-6). The iliofemoral ligament is one of the strongest and most important in the body. In a standing position, the iliofemoral ligament receives the tension of the bodyweight and settles into an orientation that prevents falling backward and does so without the requirement of muscular activity (figure 11-7). Essentially the upper body is "hanging" on these two ligaments. This represents an exception to the rule of thumb that we do not want to engage our ligaments, as ligaments represent the last defense against injury. But in this case, the iliofemoral has, over the course of human history, adapted to tolerate constant engagement. This anatomical arrangement also provides a pretty solid argument against the position taken by doctors, nurses, assembly line workers, or anyone spending lots of time on their feet during the workday - that they are on their feet most of the day, so they get enough exercise at work. There is very little additional energy expenditure from standing and working compared to sitting and working. It is interesting to note that in a study of medical professionals, as case load increased, fitness decreased. Essentially as case load increased, there was more time spent standing at bedside or charting and less moving.

The fourth extracapsular ligament is called the zona obicularis and forms a band-like structure around the capsule of the acetabulum. Think of a drum head with the femoral neck protruding through it and being held in place with a rubber band (figure 11-8). It creates a retention barrier for the femoral head, keeping it in close contact with the interior of the acetabulum

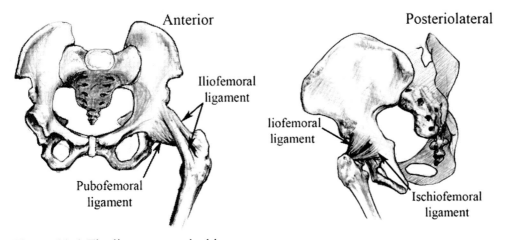

Figure 11-6. The ligaments at the hip.

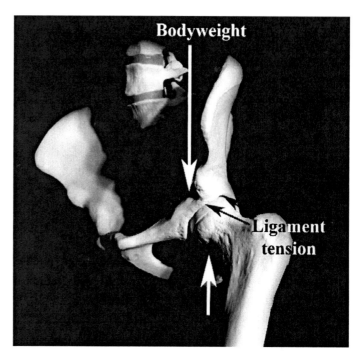

Figure 11-7. The weight of the trunk is transmitted through the pelvis to the acetabulum. A counter-force is transmitted up through the leg to the femoral head. The iliofemoral ligaments connects the two segments of the system. The end result is a balanced force across the hip - postural position maintained with only minor accessory muscular activity.

.

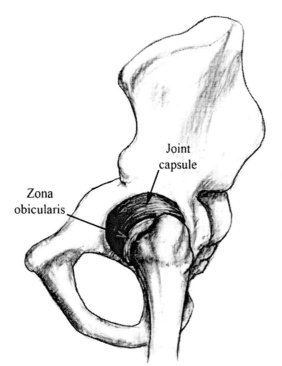

Figure 11-8. The zona obicularis.

The single intracapsular ligament, present inside the acetabulum, is the ligamentum teres. It is attached at the acetabular notch inside the bowl of the acetabulum and attaches to the head of the femur (figure 11-9). This ligament is a defense, but not only against joint. It is primarily a defensive structure for the arterial circulation to the femur, with its tough connective tissue guarding against bony contacts and impingement.

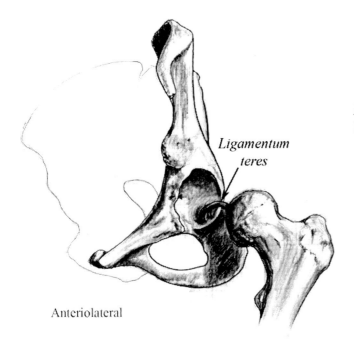

Figure 11-9. The ligamentum teres.

Ligamentum teres

Anteriolateral

MUSCLES

Gluteus medius – The gluteus medius attaches proximally to the ilium along but just inferior to the iliac crest. It also attaches to the gluteal aponeurosis. It attaches distally to the lateral surface of the greater trochanter of the femur. The muscle belly is covered partially by the gluteus maximus (posterior third) and partially by the gluteal aponeurosis, so it is not truly palpable or visible (figure 11-10).

The gluteus medius acts to move the femur away from midline – abduction. This occurs if the pelvis is the most stable structure. If the foot is planted, the primary function becomes postural. Both of these functions are active during ambulation (walk, jog, run, etc) as the gluteus medius and gluteus minumus assist in the support of the body while on one leg thus prevent the pelvis from dropping to the opposite side and interfering with the unsupported legs movement.

An interesting observation of the gluteus medius' tertiary functions is that it can act as either an external rotator or an internal rotator of the femur. With a flexed hip, it is an external rotator. With an extended hip, it is an internal rotator. A simple physical function of a broad muscle with multiple attachments.

Gluteus minimus – The smallest of the gluteal group, the gluteus minimus is a fan-shaped muscle, attaching proximally to the outer surface of the ilium well below the iliac crest. At the base of the fan is a tendon that attaches distally to the anterior portion of the greater trochanter of the femur (figure 11-10).
Due to its diminutive size, the gluteus minimus is a weak abductor of the femur or, if the foot is planted and stable, it is a postural stabilizer of the hip.

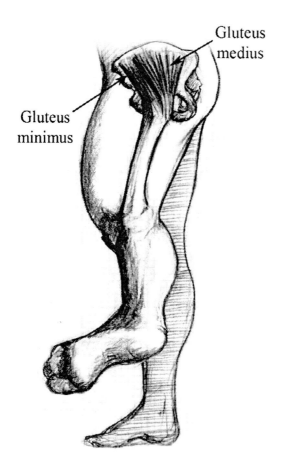

Gluteus
medius

Gluteus
minimus

Figure 11-10. The gluteus medius and gluteus minimus.

Tensor fascia latae – This odd little muscle with an exceedingly long tendon attaches proximally on the outer lip of the iliac crest and from the anterior superior iliac spine (figure 11-11). The muscle is just a few inches long and its tendon becomes insinuated within the two layers of the iliotibial band or tract along the middle and upper third of the thigh. The term fascia latae refers to a sheath of connective tissue investing the leg from the sacrum down to the knee. The iliotibial band or tract portion of the fascia latae terminates at the lateral epicondyle of the femur and the head of the fibula.

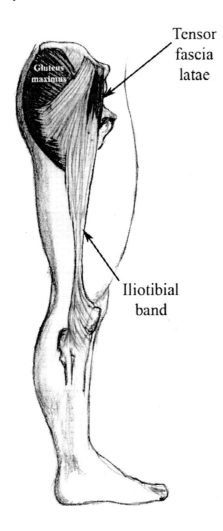

Figure 11-11. The tensor fascia latae.

When the pelvis is the most stable structure, the oblique arrangement of its fibers enables it to abduct the thigh, but owing to the muscles very small size it does not contribute much. It can also assist with internal rotation and flexion of the hip. If the foot is planted, as in normal standing posture, the tensor fascia latae carries out the task it is named for - a "tensor" of the fascia latae. It adds

145

tension to the connective tissue surrounding the gluteal group and thigh. In this role it will help stabilize the pelvis as it rests on the head of the femur. Its attachment to the iliotibial band allows it to also stabilize the articulation of the femur and the tibia.

Adductor brevis – The adductor brevis does exactly what the name implies, it adducts the leg. It is, as the name also implies, small. It attaches proximally to the pubis between the gracilis and obturator externus and it is posterior to the pectineus and adductor longus. It attaches distally on the femur from the lesser trochanter downward along the linea aspera for a short distance (figure 11-12). This is a deep muscle and is not palpable.

Adductor longus – The adductor longus attaches proximally to the body and inferior surface of pubis. Its fibers are oriented to course laterally to its distal attachment on the middle third of the linea aspera of the femur (figure 11-12). As the name implies, it is an adductor of the femur but if you look at its two attachment sites you can see that it can also assist in medial rotation of the leg.

Adductor magnus – The adductor magnus is the largest and most powerful of the adductors of the femur. As with the other adductors it is located on medial side of the femur. Sometimes the muscle is considered to have two discrete segments, the adductor segment and the hamstring segment. The adductor segment attaching proximally on the inferior side of the pubis and the hamstring segment attaching to the ischial tuberosity. Such a discrimination is probably unwarranted as the adductor magnus is not part of the hamstrings, it is only close to them. The muscle attaches distally to the lower half of the femur along the linea aspera terminating at the medial epicondyle (figure 11-12). Its orientation, anterio-posterior, makes it a mild contributor to hip extension when standing – more so when rising from a deep squat to standing (other adductors do this too, this is just the one with the biggest impact).

There is one other architectural feature of the adductor magnus making it unique among the adductor group. There are arched openings in the muscle along the femur called osseoaponeurotic openings. Femoral circulatory vessels pass through these openings.

Pectineus – This is the most anterior of the adductors of the femur. It attaches proximally at the pubis and then the muscle is oriented to the posterior, inferior, and laterally until it reaches its distal attachment on the lesser trochanter and linea aspera of the femur (figure 11-12). It is primarily an adductor but its orientation also can support medial rotation of the femur and hip flexion. This is a very small strap-like muscle and like other small muscles, it is not a primary mover rather it is much more active in a posture maintenance role when the foot is stabile.

Gracilis – The gracilis is a long superficial muscle on the medial side of the femur. It crosses two joints in its traverse from a proximal attachment near the pubic symphasis down to its distal attachment on the inferior surface of the medial tibial condyle (figure 11-12). Its adductor function is quite apparent by its location. It also is a contributor to hip flexion and is a weak contributor to knee flexion.

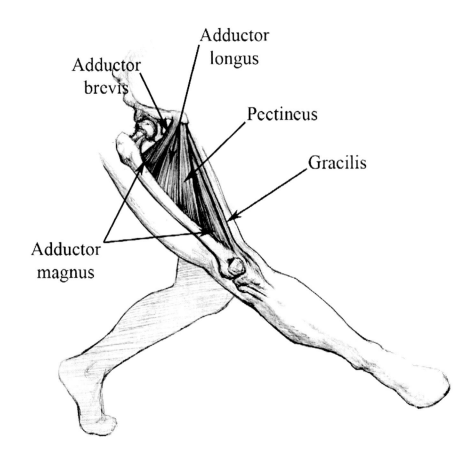

Figure 11-12. The adductors of the hip.

Iliacus – The iliacus is a flat and roughly triangular shaped muscle that lies in the concavity of the inner surface of the ilium (the iliac fossa). This makes its name fairly easy to associate with the location. As indicated above, the iliacus attaches proximally to the interior of pelvis, broadly across the ilium and forward to the anterior inferior iliac spine. It attaches distally on the femur at the

147

lesser trochanter (figure 11-13). The muscle is generally considered an anterior flexor of the hip – when the foot is the least stable structure. However, the iliacus functions as a postural muscle when the feet are planted and the body erect, pulling the top of the pelvis around the axis of the hip. If you are lying with your legs straight and sit up, the iliacus assists in sitting up. If you bend your knees while lying down, you have pre-shortened the iliacus and it can no longer contribute to sitting up. It can also contribute to lateral rotation of the femur.

Psoas major – The psoas major is another deep muscle in close proximity to the iliacus, in fact they are frequently referred to as the iliopsoas. They are indeed separate muscles and in fact in about half of the population there is a psoas major and a psoas minor. Just to state a personal preference here, pronounce psoas like you would pronounce "soap" with an "s" at the end instead of a "p". Do not pronounce it as is frequently the vogue, as two syllables – "so-ass". The psoas attaches proximally along the transverse processes and bodies of the first through fifth lumbar vertebrae. It attaches distally, along with the iliacus, to the lesser trochanter of the femur (figure 11-13). The muscle can be used as a postural muscle to maintain the lordotic arch if the pelvis or feet are the most stable portion of the system. If the feet are the least stable then the psoas becomes an anterior hip flexor. Again, as with the iliacus, if you are lying with your legs straight and sit up, the psoas assists in sitting up. If you bend your knees while lying down, you have pre-shortened the psoas and it can no longer contribute to sitting up. Conventional wisdom boldly states that sitting up with straight legs is hazardous to back health and that only the knees bent version is safe and effective. Interestingly there is no data to substantiate this claim. Some "authorities" suggest that hanging knee ups are also safer than standard sit-ups, which defies logic, as in such an exercise, the psoas (and iliacus) are quite well recruited. Think about it. Lying down with knees bent, the hips are in flexion. Hanging abdominal exercises, even with knees bent, begin with extended (straight) hips.

Strengthening the psoas major and iliacus will not and has not, in any recorded scientific or medical literature, caused vertebral deformity nor has performing old school sit ups. The exercise is perfectly safe, the psoas and iliacus are not prime movers during its execution. Its application does require a bit of sanity and not doing endless repetitions which can cause muscle soreness.

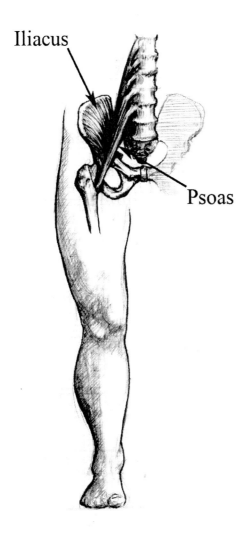

Iliacus

Psoas

Figure 11-13. The iliacus and psoas major.

Rectus femoris – As referred to in the knee chapter, the rectus femoris is located along the anterior surface of the thigh. It has a central line of connective tissue along its length and its muscle fibers radiate from it in a bipennate manner. The muscle attaches proximally at the anterior iliac spine and near the anterior lip of the acetabulum. For the purpose of hip movement, the attachment to the ilium is important. The muscle crosses the hip joint and the knee and attaches distally at the superior border of the patella (figure 11-14). The rectus femoris is the only member of the quadriceps muscle group involved in hip flexion, as the other three members have proximal attachments on the femur below the hip. The rectus femoris works simultaneously with the iliacus, psoas major, and tensor fasciae latae in hip flexion. One does need to remember gravity in movement. When you squat down, these muscles do not actively contract to pull the body down, gravity does that. Rather the rectus femoris and the other flexors act eccentrically to control and stabilize the movement.

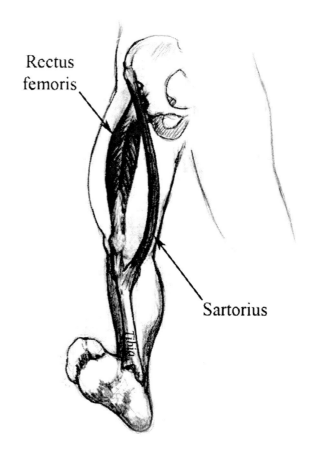

Figure 11-14. The rectus femoris and sartorius.

Sartorius – The sartorius muscle is the longest muscle in the human, obliquely spanning the distance from its proximal attachment on the anterior superior iliac spine, around the posterior surface of the medial condyle of the femur, and down to its distal attachment on the anterior medial surface of the tibial head (figure 11-14). In the final few inches of the tendon, where it curves to anterior, it merges with the tendons of the gracilis and semitendinosus to form a connective tissue structure referred to as the pes anserinus. The sartorius assists in four possible movements - flexion, abduction, lateral rotation, all at the hip, then also in flexion of the knee. If you pick up your foot, bend your knee, and move the sole of the foot to a position where you can see it (like checking to see if you stepped in something stinky), the sartorius would be carrying out all four of its functions.

Gluteus maximus - The gluteus maximus is the largest and most superficial muscle of the gluteal group (of three). In most common conversation they will be referred to as the glutes, butt, rear end, or other colorful aphorism. The muscle attaches proximally along the inner and upper ilium at the crest (iliac crest or iliac spine). It also attaches proximally to the posterior surface of the lower sacrum and to the side of the coccyx. There are also attachments to the lumbodorsal fascia and the gluteal aponeurosis. The muscle fibers are oriented obliquely - from the ilium and sacrum down to the distal attachment on the posterior femur, just distal to the greater trochanter and iliotibial tract (or iliotibial band) (figure 11-15).

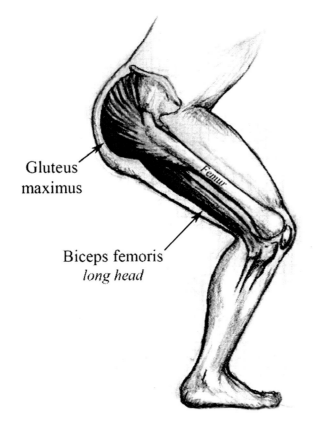

Gluteus maximus

Biceps femoris *long head*

Figure 11-15. The gluteus maximus and biceps femoris long head.

When considering the actions driven by the gluteus maximus, the most stable structure again is relevant. When the pelvis is the most stable structure (when the foot is elevated off the ground) the muscle will extend the femur to the rear. If the mass of the body is anchored below the gluteus maximus, the muscle will act on the pelvis, either supporting it and the trunk posturally or it will pull the pelvis to the posterior. The muscle's most powerful action is as a primary mover

151

in standing up from a squatting position, a movement that also recruits the biceps femoris, semitendinosus, semimembranosus, and adductor magnus. The gluteus maximus also, by virtue of its orientation and attachments, works as a tensor of the fascia latae and iliotibial band thus assists in stabilization of the femur on the tibia during standing. The lower portion of the muscle can also contribute to adduction and external rotation of the femur (and leg).

Biceps femoris – As the name suggests, this is a two part muscle located on the posterior and lateral aspect of the femur. The first part, the long head, attaches proximally to the ischial tuberosity. The second part, the short head, attaches proximally to the lateral side of the linea aspera of the femur near the proximal attachment of gluteus maximus. The muscle merges and then attaches distally on the lateral surface of the fibular head with a small strip of tendon attaching to the lateral tibia (figure 11-16). Only the long head, attaching to the pelvis, acts to move the hip. It is active in extension of the femur if the foot is the least stable structure. When the feet are stable, the biceps femoris, semitendinosus, and the semimembranosis (the hamstrings) help control back angle - when bending at the hip and during squatting.

Semitendinosus – The semitendinosus attaches proximally at the ishial tuberosity. This is an attachment and tendon it shares with the biceps femoris. In fact, for approximately the first 2 or 3 inches of both their lengths the muscles are intimately associated. The muscle ends approximately half way down the femur then transitions to a long thin tendon that attaches distally and the medial and anterior upper surface of the tibia (figure 11-16). Owing to its more posterior placement, the semitendinosus is classified as an extensor of the hip. As it crosses the knee joint, as noted in the previous chapter, it also contributes to knee flexion. Its moderately anterior distal attachment also allows the muscle to medially rotate the femur.

Semimembranosus – The semimembranosus has a somewhat similar story as the semitendinosis with a proximal attachment at the ischial tuberosity and a distal attachment at the medial tibial condyle (figure 11-16). The semimembranosus is medial to the semitendinosus. As with the previous muscle, the semimembranosus acts as a hip extensor and it can also medially rotate the femur when the hip is in extension.

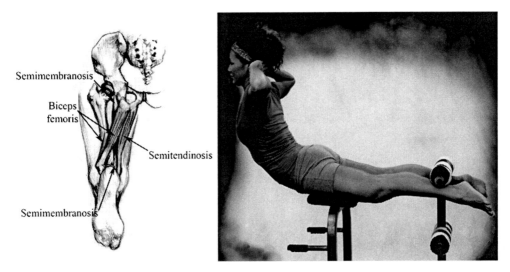

Figure 11-16. The biceps femoris, semitendinosus, and semimembranosus. The don't just extend the hip, they can assist in raising the torso during activities like the back extension when the feet are held stable (right).

The Hip's Rotator Cuff

There is a group of six small muscles that are attached proximally to the pelvis and distally to various points around the head of the femur that we can analogize to the "rotator cuff" of the shoulder (a set of muscles acting to rotate and seat the humerus into the shoulder joint). All six act to externally rotate the femur in the acetabulum and to pull the femur into close proximity with the acetabulum. They are extremely deep, not palpable, and are not especially important in terms of overt exercise performance. But the acronym for memorizing them is funny, "OOPS I Quit".

Obturator externus   Obturator internus
Piriformus   Superior gemellus
Inferior gemellus   Quadratus femoris

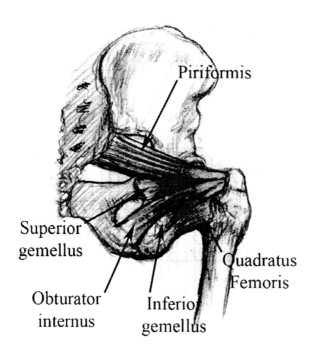

Piriformis

Superior
gemellus

Obturator
internus

Inferior
gemellus

Quadratus
Femoris

Figure 11-17. The external rotators of the hip from the posterior. The obturatur externus is not shown and crosses to the anterior of the femoral head.

"Anatomists have never been engaged in contention."

- ***William Hunter*** (1764)

12 – THE AXIAL SKELETON

The next anatomical region we will consider is the head, neck, chest and back. Although that seems like a pretty large region with many things appearing unrelated, there is one very important unifying factor – each of the bony structures to be studied is either part of, or attached solely to, the stack of bones we call the vertebral column (or much more imprecisely the spine or backbone). They are "axial" in nature as they make up the long axis of the torso and head. These bones carry out a variety of functions but the two most apparent and important functions are (1) static and dynamic structural support of the head and trunk and (2) protection of the internal viscera and the major neural structures of the body (brain and spinal cord). The axial skeleton, as it is properly known, is a very important and distinct collection of 80 bones:

Cranial Bones
Parietal (2)
Temporal (2)
Frontal (1)
Occipital (1)
Sphenoid (1)
Ethmoid (1)

Vertebral column
Cervical vertebrae (7)
Thoracic vertebrae (12)
Lumbar vertebrae (5)
Sacrum (1)
Coccyx (1)

Hyoid Bone
Hyoid bone (1)

Facial Bones
Maxilla (2)
Mandible (1)
Zygomatic (2)
Nasal (2)
Palatine (2)
Inferior nasal concha (2)
Lacrimal (2)
Vomer (1)

Chest
Sternum (1)
Ribs (24)

Auditory Ossicles
Malleus (2)
Incus (2)
Stapes (2)

THE SKULL

The skull is supported directly upon the vertebral column. It is the most superior component of the axial skeleton. It is comprised of 22 bones that can be divided into two groups; (1) bones of the cranium – those that surround the brain, (2) bones of the face – those that support the eyes nose and mouth. All of the bones of the skull, with the exception of the lower jaw, are connected at joints called sutures. These irregular, jagged, junctions are quite immovable in adults and in fact, over the lifespan, they may completely ossify and disappear. Occasionally small islets of bone, separate from the bones joined by the suture are apparent. These are called wormian bones. We will not consider all 22 bones; rather we will identify those with the greatest relevance to exercise and sport.

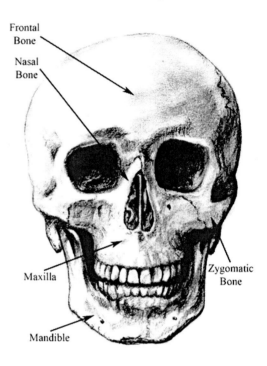

Figure 12-1. Anterior view of the skull.

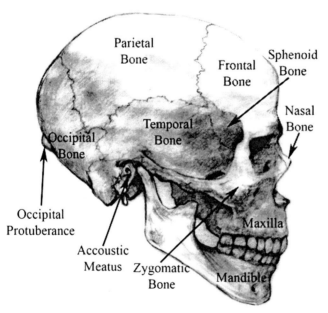

Figure 12-2. Lateral view of the skull.

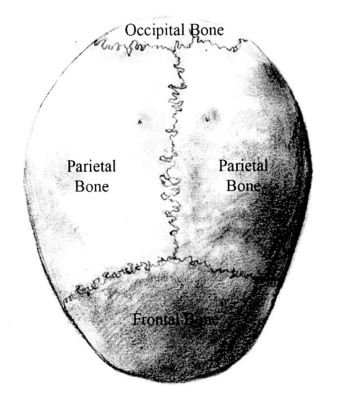

Occipital Bone

Parietal
Bone

Parietal
Bone

Frontal Bone

Figure 12-3. Superior view of the skull.

BONES OF THE SKULL

There are some bones and bony features that are easy to palpate. The position of others can only be approximated using diagrams and skeletal models, as suture joints are not palpable under the skin and subcutaneous tissues.

Frontal bone – This is the large bone that comprises the human forehead and also creates the upper ridge and the roof of the eye orbit (eye socket). It roughly occupies the area from the eyebrows up to just behind the superior hairline (this is obviously not the case with baldness). Laterally it extends from just superior and anterior to the area we refer to casually as the temple (figure 12-4). The frontal bone articulates with the parietal, nasal, ethmoid, maxilla, and zygomatic bones of each side of the skull.

Parietal bone – There are two parietal bones that together form the top and upper sides of the skull. They flank the midline of the top of the skull directly posterior to the frontal bone and extend rearward to the crown of the head (figure 12-4). Each parietal articulates with the frontal bone, the opposite parietal, the occipital, and the temporal and sphenoid bones of each side of the skull.

Occipital bone – The occipital is a roughly saucer-shaped bone at the back and lower part of the skull (figure 12-4). Approximately two-thirds down the length of the bone and medial you will find the *foramen magnum* (figure 12-5). The cranial cavity (inside of the skull) communicates with the vertebral canal (through which the spinal cord runs) via this foramen. There is an *occipital protuberance* (figure 12-2) present medially to which a number of muscles attach. This can be palpated on most people. At the level of the occipital protuberance there is a mildly raised lateral running ridge of bone called the *superior nuchal line*. It can be palpated in some individuals. A couple inches inferior to this line there is another similar line called the *inferior nuchal line*. As with the occipital protuberance, both of these features serve as sites of attachments for muscles. The occipital bone articulates with the parietal and temporal bones to the anterior, and with the inferior first cervical vertebra (the atlas).

Temporal bone – The temporal bone is located at the base and side of the skull just anterior to the occipital, inferior to the parietal, and posterior to the sphenoid bone (see next description) (figure 12-4). The posterior temporal bone contains the *accoustic meatus* (the ear hole) and just anterior to the meatus the temporal articulates with the mandible (the jawbone). This particular joint is called the *temporomandibular joint*. There are two other bony features of interest, both found adjacent to each other at the inferior andposterior part of the temporal bone behind the accoustic meatus. The *styloid process*, a small spike of bone protruding downward and the *mastoid process*, a much larger bump of bone both serve as important attachments sites for muscles. The styloid process, an attachment site for the tongue is not palpable. However the mastoid process generally is palpable by running your finger down the bone just posterior to the ear. The most inferior bump you feel is the mastoid process.

Sphenoid bone – While the sphenoid looks to be a bilateral pair of bones like the parietal and temporal bones, it is actually a single bone that spans the skull from side to side. It is one of the seven bones that articulate to create the eye orbit. The superficial aspect of the sphenoid lies just posterior to the bony ridge we call the eye socket (figure 12-4).

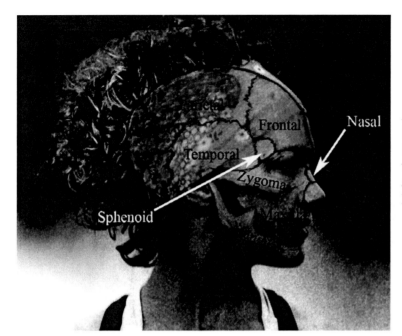

Figure 12-4. The location of the large flat bones of the skull can be approximated on a human.

Figure 12-5. The foramen magnum (big hole) on the underside of the skull.

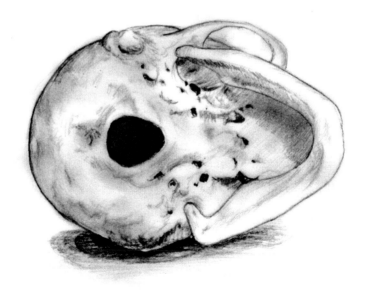

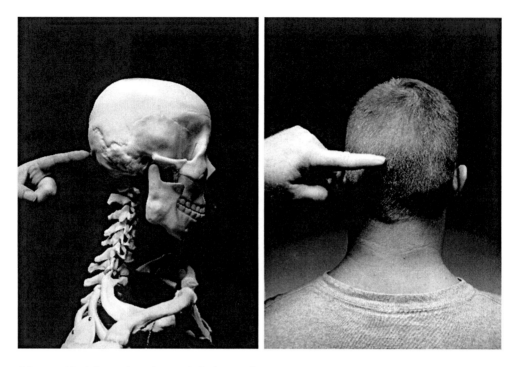

Figure 12-6. Locating the occipital protuberance.

Nasal bone – The nasal bones are two small bones that vary in size and shape in different individuals (why our noses look different). They create the bridge of the nose (your sunglasses sit on them) and flank the midline of the nose just anterior and inferior to the frontal bone (figure 12-7).

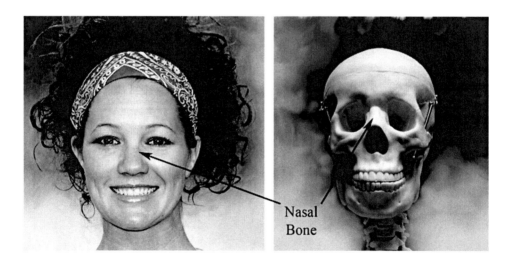

Nasal
Bone

Figure 12-7. The nasal bones can be palpated but not distinguished as individual bones.

Maxilla – The maxilla is the bone into which your upper row of teeth is set and it forms the roof of the mouth. It also creates the most medial aspects of your "cheek bones" (figure 12-8).

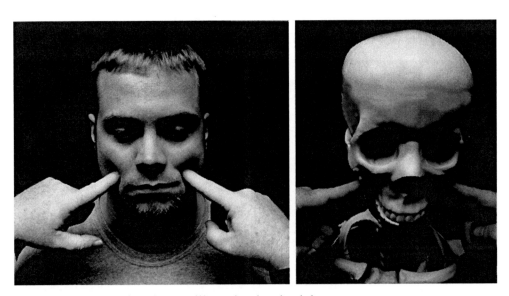

Figure 12-8. Locating the maxilla under the cheek bones.

Zygomatic bone – This bone forms the lateral and posterior component of your "cheekbones", the most curved portion. It articulates with the frontal, sphenoid, maxilla, and temporal bones (figure 12-9).

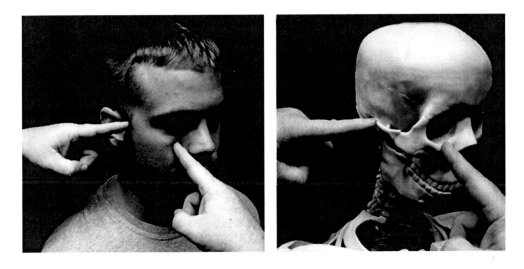

Figure 12-9. Locating the zygomatic bone.

Mandible – This is your jaw. The lower row of teeth are set into the mandible. It articulates bilaterally with the temporal bone to form the temporamandibular joints (figure 12-10). You can locate the joint by placing your fingers just anterior to the accoustic meatus then repeatedly opening and closing the mouth. This is the most freely moving and least stable bone in the skull. Not surprisingly, it is one of the most frequently fractured bones in the skull.

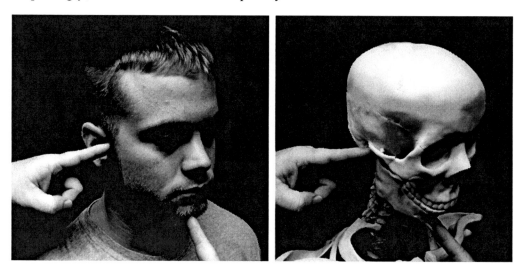

Figure 12-10. Locating the mandible.

MY BRAIN HURTS

A concussion is a fairly common occurrence in combative sports. The simplest way to think of a concussion is that the brain has been bruised. But the brain is safely tucked away inside the skull, how can it get bruised? Let's look at the basic mechanism of injury (figure 12-11). Concussions are defined as traumatically induced change in mental status with or without an associated loss of consciousness (being knocked out). They occur at a rate of about 1 per 2000 hours of sport practice and competition in combative sports. During activity the brain and skull accelerate and decelerate at identical velocities (figure 12-11a). If the skull encounters an object moving in the opposite direction, and the resulting force is large enough to stop the skull's forward motion nearly instantaneously, the brain inside the skull will continue to move forward until it impacts upon the anterior inner surface of the frontal bone (figure 12-11b). This is the injurious event that leads to concussion. After impact, the brain will "rebound" back to its normal position (figure 12-11c). In some instances the rebound motion of the brain within the skull may be vigorous enough to produce a significant collision of the brain with the interior surface of the occipital bone, a contrecoup concussion (figure 12-11d). This particular event is frequently produced by a secondary contact with an object – violent tackle leads to "B", during the fall after the tackle "C" occurs, and the combined force of the athlete and tackler striking the ground amplifies "D" and concussion occurs.

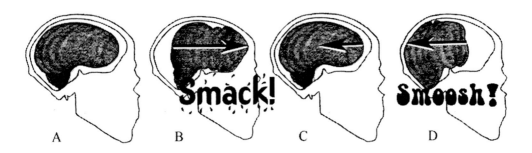

Figure 12-11. Schematic of concussion. A – Status quo. B – Rapid deceleration upon impact. C – Return to normal position. D – Post-rebound impact.

THE VERTEBRAL COLUMN

The vertebral column, a.k.a. the spine, spinal column, or backbone, is the medial and vertical axis of the body. Comprised of 33 bones each of which interlock with the bone immediately above and below them, these bones can be divided into five groups: (1) The cervical vertebrae are the smallest of the vertebrae. (2) The thoracic vertebrae are matched to each pair of ribs to form the rib cage and establish the thoracic cavity. (3) The lumbar vertebrae are the largest and thickest of the vertebrae and carry the largest load of the moveable vertebrae. (4) The sacrum forms the base of the spine and is constructed from five fused vertebrae. (5) At the very inferior tip of the sacrum is the coccyx or as some refer to it, the vestigial tail. It is formed from four or less fused bones.

A typical vertebra has three basic parts: (1) an anterior rounded chunk of bone called the body. This part of the bone is what carries the weight of the body above it. The bodies of the vertebrae become thinner as they go up the vertebral column. Vertebral bodies are separated from each other by intervertebral discs, made of tough fibrocartilage. (2) A group of posterior projections, the spinous and transverse processes, that serve as articular surfaces for adjacent vertebrae and as sites of attachment for ligaments and muscles. (3) A large central hole called the vertebral foramen (figure 12-12). The sequential stacking of the vertebrae on top of each other provides a canal through which the medulla spinalis (spinal cord) passes from its lower terminations up to its passage into the cranium and the brain.

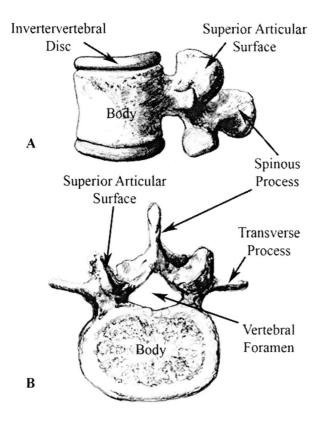

Intervertebral Disc

Superior Articular Surface

Body

A

Superior Articular Surface

Spinous Process

Figure 12-12. Basic structure of a vertebra. A – Lateral view. B – Superior view.

Transverse Process

Body

Vertebral Foramen

B

Two of the vertebrae, the *atlas* and *axis*, do not conform specifically to this generalized structure. These are the two vertebrae immediately inferior to the skull. The skull rides atop the atlas (named after the Greek mythological character who was sentenced to hold up the sky – although most sculptures show him holding up the earth) and its specific structure enables the skull to be moved forward and backwards as in shaking your head "yes". The atlas has no discernable body and no spinous process. The axis is the second vertebrae and has a unique little feature called the dens (or more precisely the odontoid process), a bony upward projection that provides a point of rotation for the superior atlas and skull. This specific structure allows the head to be rotated from side to side, like shaking your head "no".

There are five more vertebrae that along with the atlas and axis comprise the collected *cervical vertebrae*. These are the vertebrae of the neck and are located from just inferior to the skull down to just superior to the first rib. The standard abbreviation for these vertebrae, from top to bottom, is C1 (first cervical vertebrae) through C7 (seventh cervical vertebrae). On a very lean or unfit individual you may be able to palpate the spinous processes of C3 to C7 (figure 12-15).

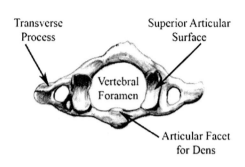

Transverse
Process

Superior Articular
Surface

Vertebral
Foramen

Articular Facet
for Dens

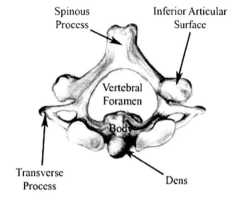

Spinous
Process

Inferior Articular
Surface

Vertebral
Foramen

Body

Transverse
Process

Dens

Figure 12-13. The Atlas.   *Figure 12-14.* The Axis.

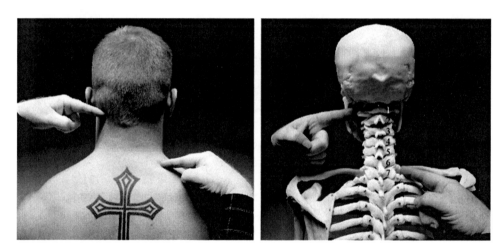

Figure 12-15. Locating the cervical vertebrae.

The next twelve vertebral bones are the *thoracic vertebrae* – the vertebrae of the thorax. These vertebrae are easy to identify as they articulate with the ribs. The first thoracic vertebrae articulates with the first pair of ribs, the second thoracic vertebrae articulates with the second pair of ribs, and so on down through the twelfth thoracic vertebrae and the twelfth pair of ribs. The accepted abbreviation for the thoracic vertebrae is T1 (first thoracic vertebrae) through T12 (twelfth thoracic vertebrae). On most individuals you can palpate the spinous processes on all twelve vertebrae (figure 12-16). With adequate pressure you can use the flanking ribs to confirm identification.

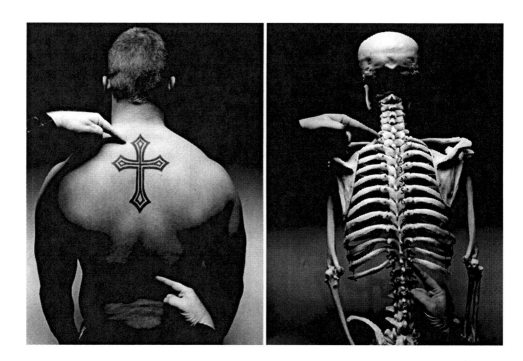

Figure 12-16. Locating the thoracic vertebrae.

The final set of vertebrae are the *lumbar vertebrae*. These are the five largest of the vertebrae. The first lumbar vertebrae, or L1 is immediately inferior to T12 and the twelfth pair of ribs. The fifth lumbar vertebrae, or L5 is immediately superior to the sacrum (figure 12-17). The *sacrum*, when viewed from the back is a roughly triangular shaped bone with four sets of foramen (sacral foramen) aligned vertically. This bone starts out life as five individual bones but fuses into one bone by about age sixteen. The bone can be palpated as it lies in between the posterior bones of the hip and just above the plumber's crack (you've all seen them) (figure 12-18). Just inferior to the sacrum is the *coccyx* or tailbone. This fusion of four or five vertebrae (individual variation) is the termination of the axial skeleton and can be palpated if desired (figure 12-19). It articulates to the sacrum by way of a fibrocartilage symphasis. This provides a degree of movement to reduce the chances of fracture.

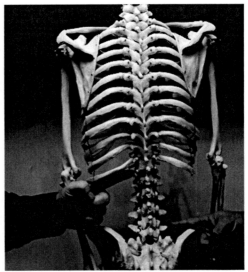

Figure 12-17. Locating the lumbar vertebrae.

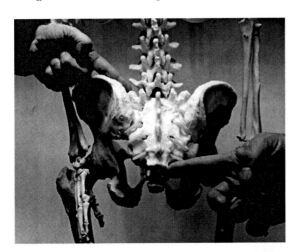

Figure 12-18. Locating the sacrum.

Figure 12-19. Locating the coccyx.

The assembled complete vertebral column has two characteristic curves to it, the kyphotic and lordotic curves. The kyphotic curve is formed by forward arching of the thoracic vertebrae. The lordotic curve is formed by the backward arching of the lumbar vertebrae. It should be understood here that a degree of kyphosis and lordosis is absolutely necessary to provide for ergonomic support for the internal viscera and the body as a whole (think of it as a structural solution to the torque that would be applied by hanging all the viscera and muscles on the anterior side of a perfectly straight axial skeleton). During sport and exercise it is important that "normal" lordotic and kyphotic curvature is maintained as excessive arching or rounding of the spinal column facilitates unequal anterior-posterior loading of the vertebral column. Such asymmetric loading can lead to injuries such as herniated discs – where the intervertebral disc is deformed by asymmetric compression to the point of rupture. From an injury prevention standpoint, maintenance of normal vertebral column curvature during any sporting activity will assist in preventing injury. From a sports performance standpoint, normal vertebral column posture is the position in which the most force can be transmitted to an opposing body – too much rounding (rolling the shoulders forward) or too much arching (pulling the shoulders back and pushing the gut forward) makes the vertebral column behave much like a shock absorber and dampens force transfer. All this does not mean that you should never train with a rounded back, in fact, it is likely quite beneficial to periodically use some rounded back exercises to prepare for the eventuality that at some point in life you will need to exert force with a rounded back – you can't lift a big rock or any other weird shaped object with a straight back.

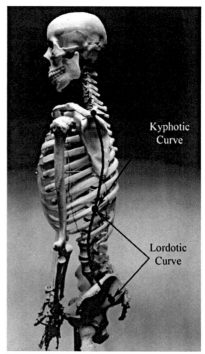

Figure 12-20. The curves of the vertebral column.

Kyphotic Curve

Lordotic Curve

Bent out of shape

Scoliosis is the presence of a lateral curvature of the vertebral column. The degree of scoliosis may vary dramatically from mild imperceptible curves to severe, physically limiting curves. Scoliosis is usually an incidental and harmless finding that presents no risk of physical limitation. However, persons with more severe scoliosis may require treatment to alleviate pain or to re-establish respiratory capacity. Treatments may include bracing, casting, or surgical correction by vertebral fusion. The onset of scoliosis development, or at least the point it is most likely to be diagnosed, is 10-15 years of age. About one third of one percent of children develop spinal curves that are considered severe enough to need treatment. There is no gender bias in its occurrence. It has an equivalent frequency of diagnosis among both genders. However, females are eight times more likely to have a progression in severity over their lifespan requiring treatment. As there is no "cure" for scoliosis there are about 6 million people living with the pathology in the United States.

It is important to understand that while scoliosis, at its most severe level of development, may be physically limiting, the vast majority of individuals with mild to moderate scoliosis can and should participate in normal physical activity. They will never perceive any physical impairment. In fact, the selection and execution of proper back strengthening exercises (posterior, anterior, and lateral) should be recommended to maintain functionality over the lifespan.

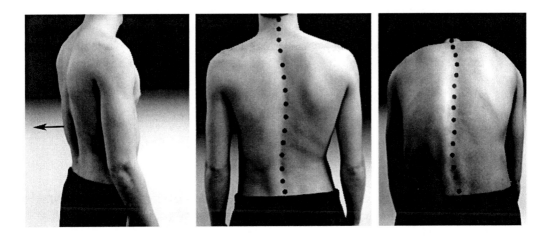

Figure 12-21. Scoliosis can be visualized by simple observation or placing marks or adhesive dots along the spinous processes of sequential vertebrae. Note that scoliosis affects lateral excursion of the vertebral column (center and right) but can also change the rotation of the vertebral column (left). This individual was not limited in athletic capacity as he played varsity basketball and trained with weights regularly.

Figure 12-22. Lamar Gant, with severe scoliosis, deadlifted 310 kg (683 lb) at a bodyweight of 60 kg (132 lb). It has been proposed that his scoliosis shortened the effective length of the lever arm formed by his torso and actually aided in his domination in the sport of powerlifting. Stand up and try to get your finger tips to touch the tops of your patella (knee cap) without bending anything other than your vertebral column. This will give you a feel for how much curvature is in Gant's vertebral column (his arms are proportional to his legs). Photographs courtesy of Mike Lambert and *Powerlifting USA* magazine.

THE CHEST

The chest, or thorax, is formed by (1) the articulations of the first seven thoracic vertebrae with the *ribs* to the posterior and the *sternum* to the anterior, (2) the articulation of the eighth, ninth, and tenth pairs of ribs with the corresponding vertebrae as with the first seven but they share a single articulation with the sternum to the anterior, and (3) the articulation of the eleventh and twelfth pairs of ribs with the corresponding vertebrae. These last five pairs of ribs are sometimes called "false ribs" or "floating ribs" as they lack attachments to the sternum. The first rib articulates with the manubrium just inferior to the clavicle (collar bone). Ribs two through five can be palpated easiest at their articulations with the sternum. For ribs six through ten, the process is to identify rib five then follow it with your fingers laterally to a point under the arm then palpate each inferior rib. For ribs eleven and twelve you will need to follow rib ten around just lateral to the vertebral column and push with a degree of force to feel the floating ribs.

The function of these bones is primarily to form the support and protective framework for all of the organs and tissues within the chest cavity and the abdominal cavity inferior to it. Although the rib cage is quite robust and impact resistant, it is also somewhat mobile. The cartilaginous joints to the anterior allow for a degree of movement that is helpful not only in external respiration (breathing) but also adds to the protective and shock absorptive nature of the ribs. The chest is quite durable and even though it has been reported that severe sneezing attacks have resulted in a fractured rib, just think of all of the boxing, mixed martial arts, rugby, and football impacts at the chest that have taken place without injury, even with direct, high force, and intentional blows.

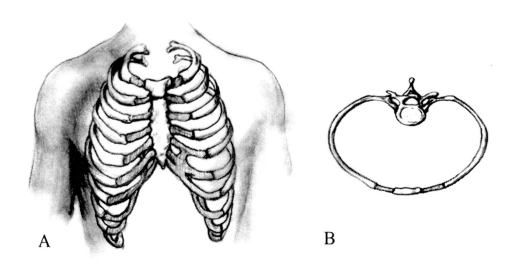

A   B

Figure 12-23. The "Rib Cage". Anterior view – left. Superior view – right.

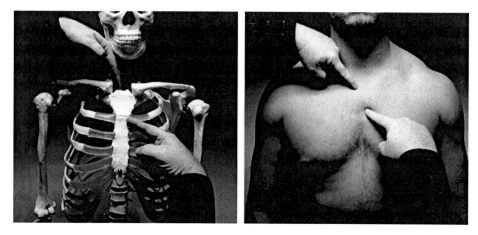

Figure 12-24. Locating the manubrium segment of the sternum.

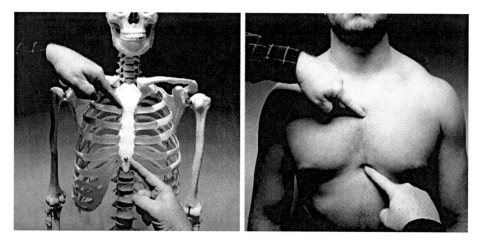

Figure 12-25. Locating the body of the sternum.

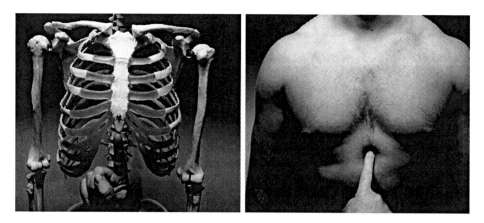

Figure 12-26. Locating the xyphoid process.

MUSCLES

The vertebral column is capable of considerable movement owing to the extensive number of joints. The cervical vertebrae display the greatest range of movement (due to the structure of the atlas and axis), the lumbar vertebrae are next in the degree of motion possible, and the least mobile of the individual vertebrae are the thoracic vertebrae (remember that 10 of them have an anterior stabilizing articulation at the sternum). The sacral and coccygeal vertebrae are fused and do not move. Combined, the vertebral column is capable of flexion, extension, lateral flexion, and rotation.

The musculature of the vertebral column occurs in pairs, specifically bilateral pairs with one of the pair present on each side of the column. This arrangement along with separate innervations allow these muscles to act both together (bilaterally) to produce symmetrical motion and individually (unilaterally) to produce asymmetric movement. For example the contraction of two muscles together may extend the neck (roll the head back) whereas contraction of only the one that is right and lateral to the vertebral column may cause rotation of the head to the right. Analysis of what vertebral muscles do is fairly simple: muscles attaching on the anterior surface of the column with cause flexion, muscles on the posterior surface cause extension. Muscles used unilaterally produces lateral flexion. Selected recruitment of opposing and complimentary individual anterior and posterior muscles will produce rotation. Often you will find that simultaneous contraction of muscle pairs does not produce any motion, rather it produces joint stabilization through isometric force production. This last observation is the primary reason that the back musculature must be strengthened in order to improve both sport participation safety and performance.

The axial skeleton serves as an attachment for a massive number of muscles. The muscles considered here are those that are directly involved in sport and exercise movement, are attached to only various segments of the axial skeleton, or in a couple instances where although the distal attachment of a muscle is outside the axial skeleton its major representative function affects the axial skeleton. There are also other muscles that attach to the axial skeleton and act upon other body parts that will be discussed in later sections of this text. You will note that there are a multitude of intrinsic vertebral muscles and other muscles not considered here (*minor movers* in table below). These are smaller muscles, frequently acting at only one of the multitude of the joints present in the axial skeleton.

Major Movers	Minor Movers
Sternocleidomastoid	Rotatores Thoracis
Scalenus	Semispinalis Thoracis
Iliocostalis Cervicis	Semispinalis Cervicis
Iliocostalis Thoracis	Semispinalis Capitus
Iliocostalis Lumborum	Splenius Cervicis
Longissimus Capitus	Longus Colli Cervicis
Longissimus Cervicis	Longus Capitus
Longissimus Thoracis	Rectus Capitus Anterior
Spinalis Capitus	Rectus Capitus Lateralis
Spinalis Cervicis	Rectus Capitus Posterior Major
Spinalis Thoracis	Rectus Capitus Posterior Minor
Psoas Major	Obliquus Capitus Inferior
Quadratus Lumborum	Obliquus Capitus Superior
Rectus Abdominus	Intertransversarii Lateralis
External Obliquus	Intertransversarii Mediales
Internal Obliquus	Interspinales
	Multifidus
	Longissimus Lumborum

Sternocleidomastoid – The sternocleidomastoid attaches to both the sternum and clavicle near the sternoclavicular joint (near the manubrium) and runs upwards and attaches to the mastoid processes just posterior to the acoustic meatus (behind the ear) and produces movements such as nodding and turning of the head. When recruited unilaterally, the sternocleidomastoid tilts the head to its own side and can rotate the head so it faces the opposite side. When recruited bilaterally, it flexes the neck or raises the sternum. This latter action is contributory to forced inspiration (voluntarily taking as big of a breath as you can). The sternocleidomastoid is quite pronounced in comic book super heroes. It will also be relatively pronounced in powerlifters who do heavy bench presses routinely, as the muscle is recruited to elevate the chest in order to shorten the distance between the chest and the arm lock-out position at the top of the movement. Wrestlers and football players will also have well-developed sternocleidomastoids as they actively train the neck musculature as part of their physical preparation for competition. To palpate this muscle, place your palm on the forehead and provide resistance to lateral flexion while simultaneously rotating the skull (try to push and rotate the head forward).

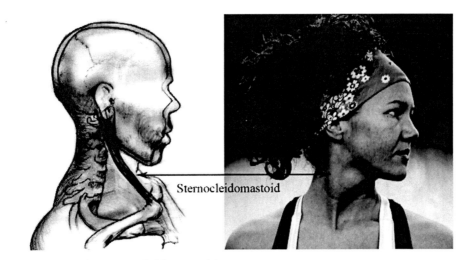

Figure 9-27. The sternocleidomastoid.

Scalenus – There are three scalenus muscle segments, the anterior, medius, and posterior. They are attached to the transverse processes of the second through seventh cervical vertebrae and the first two ribs. The scalenus anterior and medius, when contracted elevate the first rib or rotate the neck to the opposite side of the body (from the muscle). The posterior scalenus acts to elevate the second rib and tilts the neck to the same side of the body (same side as the active muscle). The scalenus lies along the anteriolateral aspect of the neck and assists in lateral flexion of the head against the resistance (tilt the head to the left or right). This is a deep muscle and is likely not palpable except in exceptional instances.

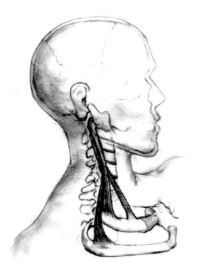

Figure 9-28. The scalenus.

175

Erector Spinae – This is not a single muscle rather the term is used to name a group of muscles that maintain normal vertebral extension (they maintain posture) and provides for extension of the vertebral column against resistance. There are three basic muscle groups;

- Iliocostalis – three defined segments; cervical, thoracic, and lumbar
- Longissimus – three defined segments; cranial, cervical, and thoracic
- Spinalis – three defined segments; cranial, cervical, and thoracic

The iliocostalis is the most lateral of the group, the longissimus is more medial, and the spinalis is the most medial. The erector spinae run the length of the vertebral column on both sides attaching to vertebral processes and posterior rib segments along they way. In a well-developed power athlete they may appear quite large, similar to having two baguettes along each side of the vertebrae. If you watch a lean individual walk from behind you can see the erectors contracting alternately with each opposing step. This aids in ambulation and in maintenance of posture. These muscles can be recruited to extend the vertebral column in part or in total. They have a heavy postural role in maintaining normal kyphotic and lordotic curves in both unloaded and loaded conditions. Strengthening this set of muscles is an important part of sports conditioning as it contributes to both orthopedic safety and to performance. There are lots of exercises and machines that "isolate" this muscle group, but simple deadlifts, Romanian deadlifts, and cleans are among the most effective developers of these muscle's function and mass. To visually inspect and palpate the erector muscles, have someone bend forward at the waist and arch his or her back (lift the chest).

Psoas major – The psoas major is not precisely an axial skeleton muscle. It is a fusiform and deep muscle that attaches proximally to the costal processes of the lumbar vertebrae, the lateral surfaces of the twelfth thoracic vertebra, and lumbar vertebrae one through four. Its distal attachment is on the lesser trochanter of the femur thus making it more properly a hip muscle. It is generally considered a hip flexor and a contributor to the forward leg swing of walking. However, if the hip is held immobile the psoas contributes to flattening or rounding of the lordotic curve, or if the erector spinae are strongly invoked the psoas can contribute to exaggerating the lordotic curve. If you can control and manipulate the arch and round in your low back, you have pretty decent control of your psoas (among many others).

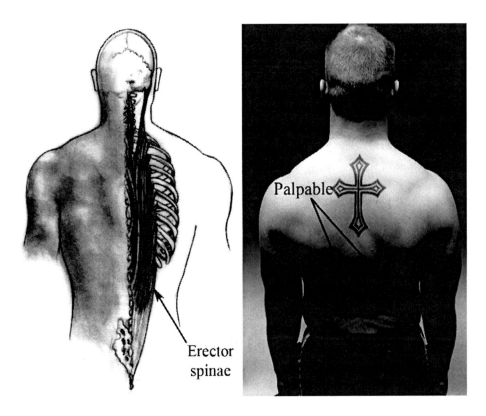

Figure 9-29. The erector spinae group of muscles.

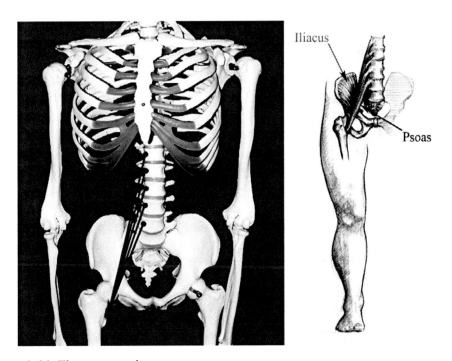

Figure 9-30. The psoas major.

As mentioned earlier, there is a bit of misinformation that has been running amuck for a number of years about this muscle and the sit-up. Sit-ups have been condemned in the popular media and academic conventional wisdom jumped on board with the condemnation based on non-existent data and logic. Here is the break down. Allegedly straight legged sit ups, and sit ups in general, are guilty of over developing the psoas muscle and inducing anterior exaggeration of the lordotic curve thus causing back pain. Hmmm. An interesting accusation that is specious (sounds good but lacks substance). The result of this rampant speculation was that the standard prescription for straight leg sit ups was to replace them with bent-knee sit ups. Elevating the femur shortens the psoas so it cannot generate force to contribute to the sit up – thereby leading to no muscle load and no development. The psoas is left flaccid and untrained and we are supposed to better off with a weaker muscle. Then crunches were invented to replace bent knee sit ups to completely remove hip flexor activity as the trunk never rotates around the hip during a crunch.

Lets use some of the same logic here on another muscle that acts on the vertebral column during a crunch, the rectus abdominis, or your six-pack muscle (which will be discussed shortly). This muscle pulls down on the anterior rib cage and flexes the vertebral column forward. Millions of people do rectus abdominis exercises every day. Do we see millions of people walking around with permanently rounded backs from crunches? No we don't. And the likelihood that strengthening the psoas during sit ups will cause an epidemic in over-arched low backs is equally as nonsensical. And there is no evidence of any merit to demonstrate that sit ups of any kind are unsafe or induce low back pain.

There is another extension of this problem area. The hanging crunch has also been prescribed by many worried about back pain in order to avoid sit up development of the psoas. OK. Just think about this from a stable structure perspective. If you are hanging and raising your knees to your chest aren't you still working your psoas? Yes you are. It is obvious that many people who influence what exercises are considered safe and effective really haven't thought about it very much.

Quadratus lumborum – The quadratus lumborum is a moderately deep four sided muscle located in the area of the lumbar vertebrae. Its contraction can produce a number of actions; lateral flexion of vertebral column if only one of the pair is recruited, extension of lumbar vertebral column with bilateral contraction, and during forced exhalation it immobilizes the twelfth rib pair. Its proximal attachment is on the iliac crest (top of the hip-bone) and it is attached distally to the lower border of the twelfth rib and the transverse processes of the upper four lumbar vertebrae.

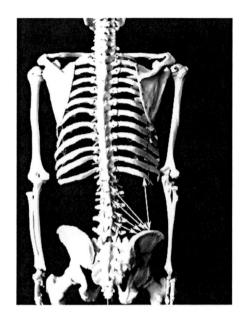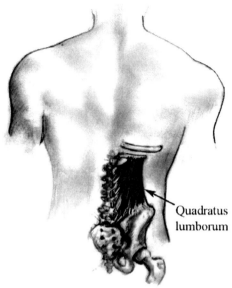

Quadratus
lumborum

Figure 9-31. The quadratus lumborum.

Rectus abdominis –The rectus abdominis muscle is a bilateral paired muscle that courses vertically on each side of the anterior wall of the abdomen. The two lateral muscles are separated by a medial band of connective tissue called the linea alba. The rectus abdominis extends from the pubic symphysis and pubic crest inferiorly to the xyphoid process and fifth through seventh costal cartilages superiorly. There are three transverse fibrous bands called tendinous inscriptions. Combined with the linea alba, these connective tissues divide the active musculature of the abdominis rectus into six roughly symmetrical segments. In very lean or very muscular individuals, the defined appearance of these segments is termed a "six pack" or of more antiquarian origin "washboard abs".

The rectus abdominis is an important anterior postural muscle responsible for maintaining a balanced isometric force (countering the erector spinae) for normal posture or actively flexing the lumbar spine. The rectus abdominis can also assist in forced respiration, as its active contraction can reduce abdominal and thoracic volumes.

 To inspect the three muscle bellies of the abdominis rectus, have you're a person lie down on his or her back. Palpate the lowest rib proximal to the xyphoid as well as the area just lateral to the naval. Have your subject elevate their head and roll their shoulders forward. You should be able to feel the tension of the muscles. Move your hand up from naval level to the xyphoid process paying attention to the location and symmetry of the individual segments.

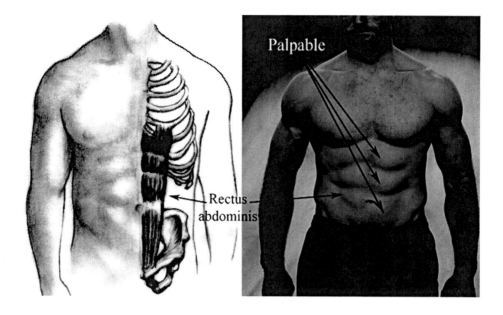

Figure 9-32. The rectus abdominis.

Does a Beer Gut have a Six-Pack?

Now is a good time to tackle one of the quintessential exercise issues faced by trainees, trainers, and coaches. You, like thousands of your peers, want a six-pack rectus abdominis. Every day you do hundreds of crunches, twists, or sit-ups, but that desired definition just never pops up. You can feel the muscles under your skin and you're even pretty lean, but your abs just don't show. What's the problem? It's in your programming. It lacks intensity. Compare the physiques of a distance runner, a sprinter, and a pro bodybuilder. Of the three, which has the most defined muscularity? Low speed activity with bodyweight begets no rip. High speed activity with bodyweight improves the amount of rip. Weighted exercise maximizes rip. The problem with your abs is that you're training them for a marathon when you actually should be training them to look big and strong. The endless bodyweight repetitions of crunches and sit-ups that most people believe will deliver a six pack will only deliver muscular endurance. Why strength-train the rest of your body and endurance-train your abs?

To fix this problem you only need make a few modifications to what you already do. First, add resistance. The rectus abdominis is a muscle, just like the quads, just like the pecs. What do we do if we want a bigger and more defined chest and thighs? Do a bunch of push ups and air squats? No. We bench and squat and we progressively add weight for lots of reps to maximize hypertrophy (increase muscle mass).

So first, we add weight to the bar – in this case we actually add weight to the upper body. Instead of the standard crunch – remember a crunch is simple spinal flexion, get a barbell plate, get in your crunch start position, hold the plate at arms length over the upper chest and chin, then do your crunch movement, as high up as you can, while maintaining the plate's position over your chest and chin. (Shove your toes under a plate rack or have a buddy hold your feet down). Think of it as punching the weight straight up in the air (figure 9-33). You can just go ahead and come all the way up and do a sit up if you want, its OK.

Now you have a weighted crunch that adds intensity to your abdominal training, a resistance that can and should be increased as you get stronger. Start with a weight that is heavy enough to make you work. Don't pick a weight you can do easily, any program that says it's an easy way to fitness is an easy way fail.

Second, we do the correct number of reps, around 10-12 reps for multiple sets. Dr. Mike Stone, former head of Sport Physiology for the United States Olympic Committee, proposes that maximal muscle growth is stimulated via 5 sets of 12 repetitions. That means a maximum of 60 repetitions in a multiple-set weighted crunch workout will deliver big results! Start with 3 sets of 12 repetitions. Add repetitions in a new set EVERY workout and when you can do 5 sets of 12 repetitions, add 5-10 pounds to your plate weight, drop back to 3 sets of 12 repetitions, and then work back up to 5 sets of 12. In short order you will be able to crunch the big 45's with no problem. At that point you do not want to start crunching a stack of plates. That can be a little risky, rather you need to get an easy-curl bar and begin to add your weight to a bar.

Cutting the bodyweight reps and progressively adding weight to your crunches, or sit-ups, will rapidly develop the rectus abdominis and will deliver the six-pack muscles. You can build awesome abs with intense training but it's up to you to make sure that everyone gets to see the results of your hard work by making sure that you don't bury the evidence under the results of high intensity eating.

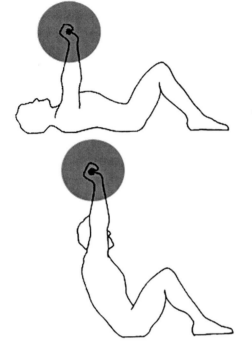

Figure 9-33. Weighted crunch.

External Obliquus – The external obliquus (commonly called the external oblique), on the lateral and anterior abdomen, is a broad, thin, and roughly quadrilateral muscle. In general the external obliquus is not visible due to subcutaneous fat deposits and the relatively small mass of the muscle. The muscle arises from eight superior points of attachment on the external and inferior borders of the fifth through twelfth ribs. The fibers of the muscle are arranged in an oblique fashion (on an angle). The muscle attaches inferiorly on the anterior half of the outer lip of the iliac crest. There is a coalescence of connective tissue (aponeurosis) at the lower border of the external obliquus muscle that forms the inguinal ligament. The muscle acts to pull the rib cage downward (flexion of the vertebral column) either bilaterally or unilaterally. This movement compresses the abdominal cavity thus increasing intra-abdominal pressure. This particular function contributes to vertebral stability during the execution of the Valsalva maneuver (attempted exhalation against a closed glottis) and it also contributes a small amount to both flexion and rotation of the vertebral column. To palpate this muscle, place your palm just below the most lateral and inferior aspect of the rib cage and alternately laterally flex their trunk (with considerable isometric force at the bottom) then extend it in the opposite direction.

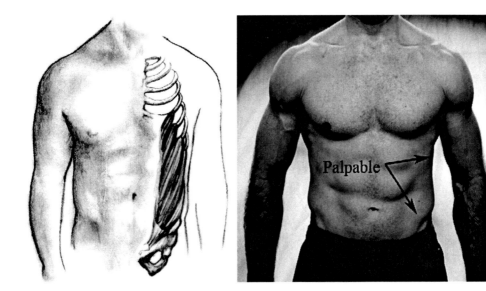

Figure 9-34. The external obliquus.

Internal obliquus – The internal obliquus muscle (commonly called the internal obliques) lies just underneath the external oblique so it is not palpable. It is quite similar in structure to the external obliques but its fibers are oriented on the diagonal but perpendicular to the external oblique muscle. Its proximal attachment is through the thoracolumbar fascia of the lower back and the anterior of the iliac crest (the top of the hip bone). The muscle attaches distally to the inferior borders of the tenth, eleventh, and twelfth ribs and across to the linea alba. The internal oblique performs two basic functions. It is an antagonist (opposing muscle) to the diaphragm, compressing the thoracic cavity to drive exhalation. It also acts to rotate and bend the trunk by pulling the rib cage and midline towards the hip and lower back, of the same side as the active muscle is located – a same side rotator.

Diaphragm – The diaphragm is a dome-shaped muscle with a substantial connective tissue component that forms a septum separating the thoracic and abdominal cavities. It's shape follows the contour of the area defined by the ring of the rib cage and the vertebral column with the most peripheral portion consisting of muscle fibers that attach to the circumference of the ribs and converge centrally to attach to a central, sheet-like tendon. One segment of the muscle arises from the back of the xyphoid process, another segment that arises from the inner surfaces of the lower six ribs, and another segment arises from the anterior lumbar vertebrae.

The diaphragm is the primary muscle of breathing (external respiration). During inhalation, when the diaphragm contracts, the central tendon is pulled downward thus increasing the volume of the thoracic cavity. This reduces intra-thoracic pressure to lower than atmospheric pressure (forms a vacuum) thus forcing air to be drawn into the lungs. When the diaphragm relaxes, the elongation of the muscle allows the central tendon to move back up, reversing the process and expelling air.

No one really talks too much about training the diaphragm. The convention is that it works 24-7-365, that is plenty of work, so why worry about anything specific? Logic tells us otherwise. Does our heart muscle not work 24-7-365 and do we not make it work harder intentionally and in specific manners to improve performance? Of course we do. The diaphragm, just like all muscles, is adaptable to stress. We need to consider working the diaphragm with progressing intensity, to improve its function – forget the "talk test" and work hard to breathe hard in order to be able to play hard.

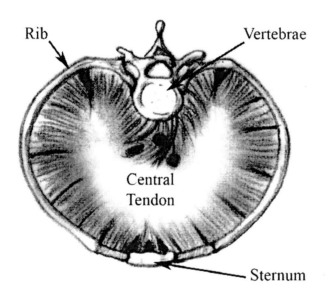

Rib

Vertebrae

Central
Tendon

Sternum

Figure 9-35. The diaphragm.

13 – THE SHOULDER

The shoulder is one of, if not the most, mobile joints in the human body. We can reach up, reach down, back, sideways, and all points in between because of its construction and extensive range of motion. We can throw, pull, push, and support loads in virtually any position thanks to its mobility and musculature. Think about any sport or exercise with which you are familiar and you will likely conclude that the shoulder is active, in some aspect either dynamically or statically, in virtually every movement in sport or fitness.

The shoulder joint is a "ball and socket" joint and is comprised of three bones, the clavicle, scapula, and the proximal end of the humerus (figure 13-1). Often when people talk about both shoulder joints along with their attachments to the axial skeleton, you might hear the term "shoulder girdle". Another commonly heard layman's term and in fact the one we have been using so far is "shoulder joint". The shoulder is not just a single joint, there are three important articulations in the region. The anatomically correct names of these three joints are the acromioclavicular, glenohumeral, and sternoclavicular joints (figure 13-2). Together, structurally and functionally, they form the freely moving joint system we call the shoulder. There is a minor drawback to such mobility, it is accompanied by relative instability and the shoulder can be injured more easily than other joints. Minor in that appropriate strengthening and movement technique reduces the chance for injury. For any coach or fitness trainer, it must be understood that strengthening the shoulder musculature, the prime movers specifically, is the most important aspect of injury prevention and of performance enhancement relative to shoulder movement.

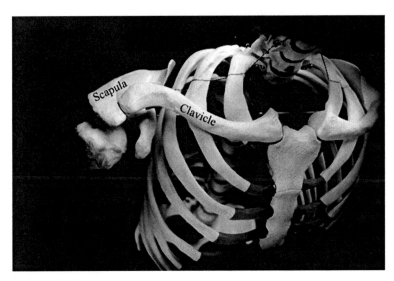

Figure 13-1. The shoulder is comprised of three bones, the humerus, scapula, and clavicle.

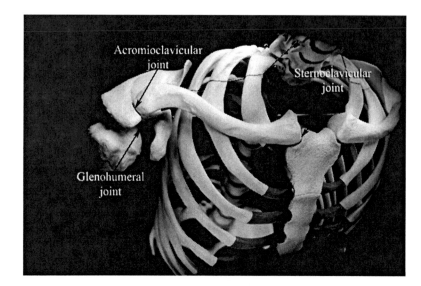

Figure 13-2. The three joints of the shoulder, the acromioclavicular, glenohumeral, and sternoclavicular.

Acromioclavicular joint

Sternoclavicular joint

Glenohumeral joint

Figure 13-3. The shoulder is capable of a large degree of mobility. The only thing saving this jukodas elbow is shoulder range of motion.

BONES

The bone of the shoulder closest to the axial skeleton is the clavicle. The common term for the clavicle is the "collar bone". You can easily remember its location as it's the bone that the front of your shirt collar rests upon. The clavicle articulates with the sternum anteriorly and to the scapula posteriorly. When viewed from the front, the clavicle has a somewhat flat appearance and forms a very shallow and long "S" shape when viewed from above. The second and most posterior bone of the shoulder is the scapula. The scapula is almost triangular in shape and is a flat bone that lies over the ribs of the upper back, lateral to the vertebral column (one scapula on each side of the vertebral column). The last bone of the shoulder set is the long bone of the upper arm, the humerus. The articulation of the humerus with the scapula is the crux of arm movement and as such, the scapula must be considered as critically important in any analysis of arm movement in sport or exercise.

With only three bones involved, the shoulder is a simple skeletal structure, however there are there are a number of important and useful to know landmarks on these bones.

Clavicle – The clavicle acts like a strut (structural support resisting a load along its length) to support the upper extremities. It also serves to protect the underlying subclavian neurovascular bundle. The complete length of the clavicle is easily palpated on anyone. It is quite superficial. The muscles attaching to it do not obscure its superior and anterior surface in front of the neck. If you look at the little dip in the line of bones across the front of the shoulders, just below the neck, the bump lateral to the suprasternal notch (the medial dip you see) is the medial end of the clavicle (figure 13-4). You can walk your fingers laterally along the length of superior surface of the clavicle until you get to the point of the shoulder where it articulates with another posterior bone of the shoulder joint, the scapula. The clavicle is a critical bone in shoulder and arm function, a fracture is quite painful and frequently renders the arm on the fractured side non-functional. About five percent of all fractures diagnosed and treated in the US are clavicular fractures. The most common mechanism of injury involves a direct and large magnitude force applied to the lateral aspect of the shoulder as a result of a fall or contact in sport (or car accident). A historical and relatively gruesome military tactic for preventing prisoner resistance in forward hostile environments is to fracture both scapulae of the prisoner, thereby preventing any aggressive prisoner actions with the upper body but maintaining their lower body mobility.

Scapula – Parts of the scapula are readily palpable depending on the degree of musculature of the individual palpated. It is much easier to palpate the scapula on very slightly built endurance runners and dancers than it is on heavily muscled strength and power athletes. The most prominent feature of the scapula is the spine of the scapula, or the scapular spine. It is a posterior projection off of the superficial face of the scapula, about 2/3's of the way up its length. Most people will refer to this feature as the "shoulder blade". The spine will feel like a bony ridge that runs obliquely, rising at a shallow angle as it crosses the upper back, medial to lateral (figure 13-6). As will be the case in most large boney features, several muscles will attach to the spine. The large trapezius muscle attaches on the superior surface of the spine and prevents palpation of the upper medial border and the superior border of the scapula. Palpation of the spine out to the most lateral (furthest out) portion of the scapular spine will take you to the point of the shoulder. This structure is the *acromion process*, the flattened terminal portion of the scapular spine (figure 13-7). This area should be familiar as the acromion process articulates with the clavicle to form the *acromioclavicular joint*.

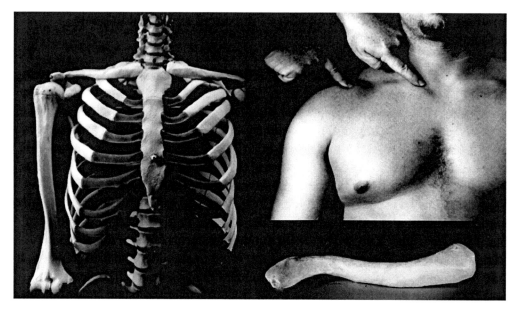

Figure 13-4. The clavicle in place on a skeleton (left), *in situ* in a human (right top), and separate (left bottom).

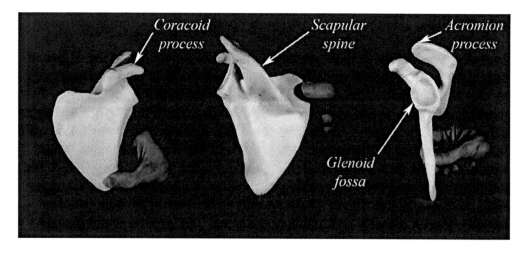

Figure 13-5. The scapula in three views.

Below the scapular spine you will find the scapular borders. A *lateral and a medial border* of the scapula can be identified on lean individuals by pressing the fingers firmly into the tissues just below the medial and lateral ends of the spine and walking the fingers down the edges of the scapula (use both hands). You should see that the medial border is more vertical than the lateral. The bottom of the scapula, where the borders meet is called "inferior angle" (figure 13-8).

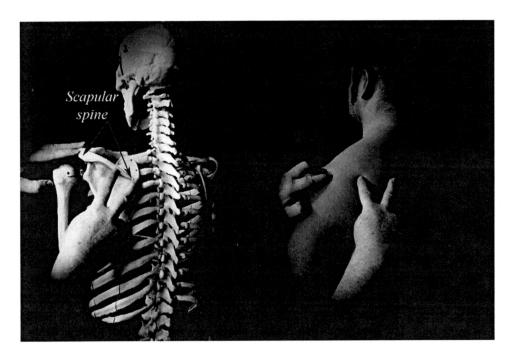

Figure 13-6. Locating and palpating the scapular spine.

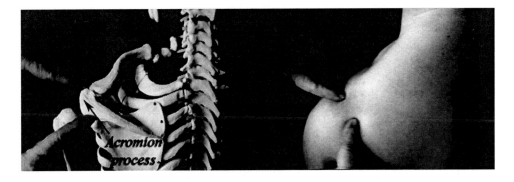

Figure 13-7. Locating and palpating the acromion process.

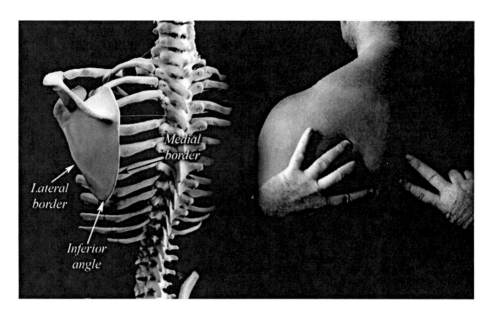

Figure 13-8. Locating and palpating the scapular borders and inferior angle.

There is another important process present on the scapula, the *coracoid process*. It is a moderately sized projection on the anterior and inferior side of the scapula and can be palpated on the anterior side of the body (figure 13-9). Stand in front of someone, place the palm of your hand (approximately) on the top of the shoulder. Wrap your thumb down to the front of the shoulder to a point in the middle of the fossa or "hollow" that appears below the clavicle and between the arm and chest. By applying firm pressure with your thumb and having your subject move his shoulder *slightly* forward and backward, you should feel a bump that moves when the scapula moves. This palpation may possible cause a small degree of discomfort.

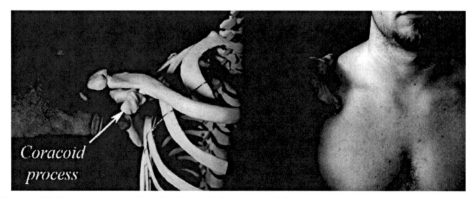

Figure 13-9. Location and palpation of the coracoid process.

Just inferior to the scapular spine along the lateral border there is a shallow, somewhat bowl shaped oval depression called the *glenoid fossa* (figure 13-5). The fossa points laterally and slightly forward and articulates with the head of the humerus. This is properly the socket of the ball and socket joint of the shoulder. In fact, glene is the Latin for "socket".

Humerus – The long bone of the upper arm is the humerus (figure 13-10). It is a relatively infamous bone, being one of the critical elements – the crossed bones – of the original "Jolly Roger" pirate flags of the late 17th century. Simple visual inspection tells you, yes, this is a long bone as it runs from the acromion down to the point of the elbow joint. Much of the bone is easily palpable but as with the scapula, the small features of the humerus may be less or more difficult to palpate and identify depending on the degree of individual muscularity.

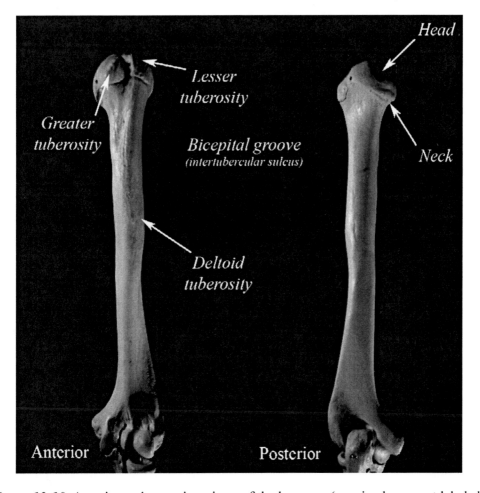

Figure 13-10. Anterior and posterion views of the humerus (proximal segment labeled).

The humerus possesses a proximal rounded head with a narrowed neck and two tuberosities. The greater tuberosity is the superior process appearing just lateral to the humeral head. The lesser tuberosity is slightly inferior, anterior and medial to the greater tuberosity. It is also palpable. A relevant landmark for locating the lesser tuberosity is the coracoid process. The tuberosity lies approximately and inch or two lateral to the coracoid process on the anterior surface of the upper arm at the shoulder. It is also anterior and inferior to the acromion. Rotation of the relaxed arm (hanging by the side) inward and outward several times will feel as a hard bump moving left to right under the skin and deltoid muscle (figure 13-11). Both the greater and lesser tuberosities are important sites of shoulder muscle attachment. The distal end, discussed in more detail in the following chapter, features two epicondyles, two processes (trochlea and capitulum), and three fossae (radial fossa, coronoid fossa, and olecranon fossa). Its body, or shaft, is roughly rounded towards the proximal, upper, portion then transitions to a more three-edged prismatic shape distally.

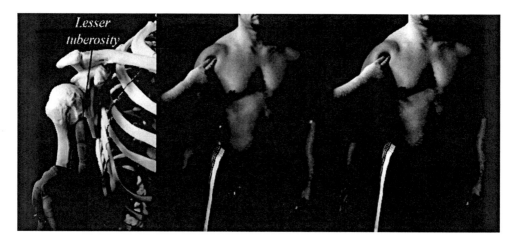

Figure 13-11. Location and palpation of the lesser tuberosity.

Another tuberosity relevant to the shoulder is the deltoid tuberosity, occurring along the lateral aspect of the humeral shaft. This low rising process will vary in degree of prominence on different individuals and heavy musculature may also make palpation difficult. On a skeletal model it is a moderately sized raised area that appears to be a little rough (figure 13-12). This makes it an excellent representation of form facilitating function. Here, the elevation and rough surface texture provides an excellent means for muscle attachment. This bony feature is located at the bottom of the deltoidus muscle. Simply trace the line of the triangle-shaped deltoid muscle (shoulder muscle) down to its narrow inferior and lateral terminus on the humerus to a lateral point between the anterior biceps brachii and posterior triceps brachii muscles (bi's and tri's to the layman). This palpation may require a bit of pressure. A relaxed arm is important for palpation as contraction of any of the three muscle groups involved will prevent palpation with low manual pressure. The more muscle tension present, the harder one has to press to feel the process.

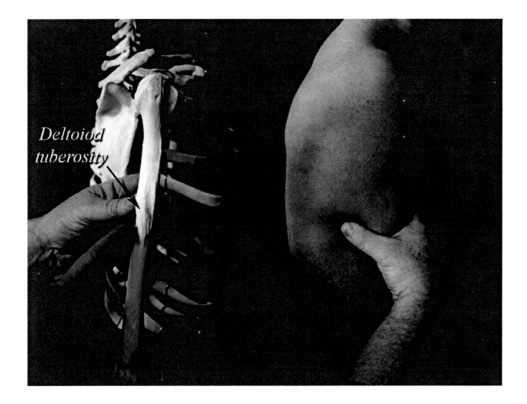

Figure 13-12. Location and palpation of the deltoid tuberosity.

It is important to remember here, as it has been throughout this text, that the ability to locate bony features helps in identification of which muscles are active in a movement or sometimes more importantly, which muscles are sore or injured. Muscles attach to bones, knowing where those attachments are enables a savvy coach or fitness trainer to rapidly deduce and solve training problems.

JOINTS

Due to the degree of mobility in the shoulder, joints and connective tissue is fairly important here. There are two divisions of cartilage present, the first is the *articular cartilage* lining the end of the bones comprising the three joints (as discussed in chapter 6). This allows the bones to glide and move on each other with little resistance. Degradation of articular cartilage (wearing out from long term use or disease processes), generally causes the joint to become painful and stiff, leading to reduced range of motion and functional capacity. The basic term for this condition is arthritis. The second division of cartilage found here is the *labrum*. This cartilage is much more rigid than articular cartilage on the ends of bones. The labrum is a ring of cartilage that functionally deepens the socket structure (the glenoid fossa) of the scapula which accepts the head of the humerus.

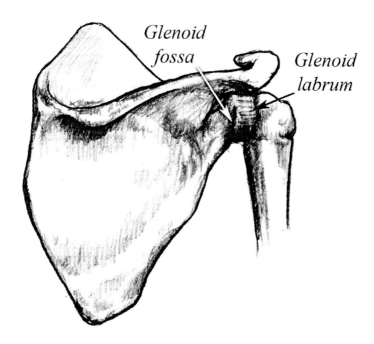

Figure 13-13. The glenoid labrum.

The articulation of the sternum and the clavicle is logically called the *sternoclavicular joint*. It is commonly thought that this joint is quite static, but it can move. Place a thumb on the manubrium, lateral to the suprasternal notch, and a finger somewhere along the length of clavicle. Movement of the shoulder anterior-posterior or superior-inferior through the complete range of motion will demonstrate the mobility of the clavicle relative to the manubrium. While this joint is mobile it is, in fact, quite stable with sternoclavicular dislocation being quite rare (it can occur in contact sport from by direct trauma). There is a joint capsule present along with a disc of cartilage within the capsule that contributes to joint stability. The *costoclavicular ligament* along with the *sternoclavicular ligament* stabilize the joint (figure 13-14). The sternoclavicular ligament crosses the sternoclavicular joint from anterior and posterior. Palpation of the ligament is possible but involves being able to discriminate between the feel of two adjacent bones and the short ligament connecting them. The sensation from ligament palpation differs from the bone in that it has some "give" to it.

The *acromioclavicular joint*, as the nomenclature infers, is the articulation between the acromion process and the lateral and distal end of the clavicle. Frequently referred to as the AC joint, it is the joint at the top of the shoulder. The architecture of the acromioclavicular joint provides for a relatively unstable joint, more so than the sternoclavicular joint. This is not detrimental in any way given the high mobility requirements of shoulder joint for function. It is possible in some individuals to find the *acromioclavicular ligament* with careful palpation (figure 13-15). To find this joint, walk the fingers of one had laterally along the superior surface of the clavicle while walking the fingers of the other hand laterally along the scapular spine. Both hands should meet at the joint where the lateral end of the clavicle and acromion process articulate. The ligament lies between the two bones. There last major ligaments of interest, the coracromial and trapezoid ligaments, connect the coracoid process to the clavicle (figure 13-15) and provides further stability for the shoulder. They are not palpable.

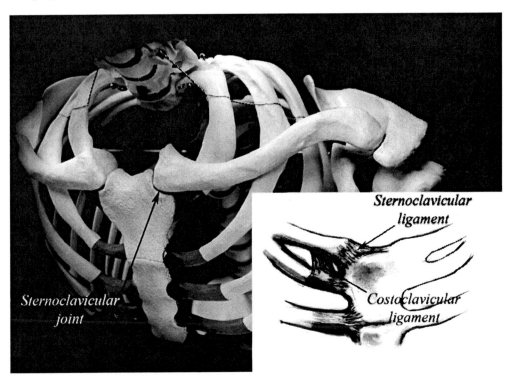

Figure 13-14. Ligaments of the sternoclavicular joint.

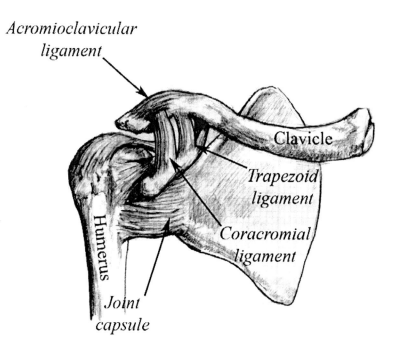

Acromioclavicular ligament

Clavicle

Trapezoid ligament

Coracromial ligament

Humerus

Joint capsule

Figure 13-15. Major ligaments of the shoulder.

The *glenohumeral joint* is the primary joint of the shoulder (figure 13-17). The head of the humerus and the glenoid fossa articulate in the characteristic ball and socket fashion, allowing the arm to rotate in virtually any anatomical plane. The shallow depth of the glenoid fossa and relatively lax connective tissue connections between the shoulder bones and the rest of the body enables the arms great mobility. Relative laxity and instability means that the glenohumeral joint, even with the presence of the labrum, is much easier to dislocate than most joints in the body. Dislocations are somewhat common, with a little less than 2% of the population experiencing at least one over the lifespan. Dislocations are more prevalent among adult males (9:1 male:female) up to about age 60, when there is a reversal, with women experiencing more dislocations thereafter (3:1 female:male). Shoulder dislocations have been documented throughout human history, documented in Egyptian tomb murals dated as early as 3000 BC.

Therapeutic correction (putting the head of the humerus back into the glenoid fossa) was documented by Hippocrates, who's technique was quite similar to the technique in fashion in that late 19th and early 20th centuries. One of the most common sport and exercise circumstances for shoulder dislocations results from violent abduction, extension, and external rotation (such as in a volleyball

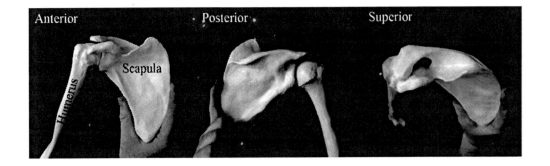

Figure 13-17. Three views of the glenohumeral joint, the articulation of the scapula and the humerus.

spike). More common in non-sporting environments, but still common in sports, is dislocation from landing on an outstretched and stiffened arm and hand after a fall. Some individuals with a history of frequent dislocation may become adept at self-reseating the humerus into the glenoid. A televised instance of this occurred in 2009 during "Nitro Circus" where Travis Pastrana crashed on his motorcycle during a novel and failed stunt. Landing inverted, his left humeral head was dislocated to the anterior. He used his right arm and body weight to wrench the humerus back into place and continued with repeated attempts to land the stunt. Most individuals will require medical attention with dislocation. There is a soft tissue capsule that encircles the glenohumeral joint, attaching to the scapula and the humerus. The capsule is also contiguous with the superior biceps brachii tendon. The capsule is lined by a synovial membrane and is reinforced by the ligamentous structures of the three joints present in the region. There is another ligament called *semicirculare humeri* crossing transversely between the posterior sides of the tuberosities of the humerus, further strengthening joint capsule.

MUSCLES

The muscles of the shoulder joints are arguably the most important component of the musculo-skeletal system of the shoulder. They must provide for mobility for every directional movement around the ball and socket joint and they must provide for joint stability preventing dislocation, given the relative lack of bony support. Further, every load carried, pulled, or pushed with the hands transmits its weight to the axial skeleton then to the ground through the shoulder skeleton and musculature. You can't pull off the floor without a shoulder muscle contribution, you can't push overhead without a contribution, even something as simple as a holding a bag of groceries recruits the muscles of the shoulder. Muscles of the shoulder can be divided into two strata, the palpable superficial muscles and the non-palpable deep muscles.

Trapezius – The trapezius, also called the spinotrapezius, is the large and recognizeable superficial muscle of the upper back (figure 13-18). Its geometric shape is roughly trapezoidal and thus the source of the muscle's name. It extends vertically from as high as the occipital bone and occipital protuberance to as low as all of the thoracic vertebrae. It spans laterally from the vertebral processes out to the spine of the scapula. There are distinct segments of the muscle, each with specific muscle fiber orientations:

The most superior group of fibers runs from their proximal attachments on the axial skeleton downward and laterally to their distal attachment on the posterior side of the lateral border of the clavicle.

The middle group of fibers begin approximately where the fiber orientation is horizontal. They track between the medial aspect of the acromion to the superior and posterior border of the scapular spine.

The most inferior group of fibers proceeds upward and laterally from the lumbar vertebrae and converge upon the scapula at the medial end of the scapular spine.

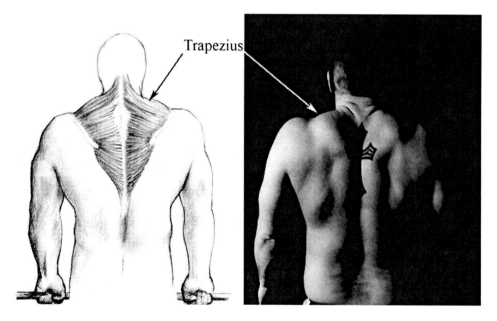

Figure 13-18. The trapezius.

The proximal and distal attachments of the trapezius are, in general, easy to palpate. An important observation here is that the vertebral column is held stable during most movements. This implies that the scapula will generally move relative to their attachments to the axial skeleton – they will move toward and away from the axial skeleton, not the other way around.

The superior trapezius elevates the scapulae and therefore the shoulders (makes you shrug). The middle trapezius pulls the scapula towards midline – retracts them. The lower trapezius draws the scapulae downward – depresses them. It is extremely common to see shoulder shrugs as the only exercise prescribed to strengthen and hypertrophy the trapezius. But as the anatomical structure of the muscle suggests, this exercise only works one of the three contractile motions of the trapezius, elevation. This is an incomplete exercise treatment. Deadlifts require a degree of all three muscular functions as do cleans and power cleans, suggesting them as superior exercises for trapezius strengthening and development. If the muscle isolationist approach to training is for some reason used in programming, complimentary exercises like rows and dips must be used along with shrugs to adequately develop the trapezius in entirety. This is a less efficient, less functional, and more time consuming methodology.

Deltoid – The deltoid is a moderately large, delta shaped muscle lying over the glenohumeral joint. The muscle attaches to several skeletal features and each attachment is associated with a distinct segment of the muscle. The anterior segment of the muscle proximally attaches at the inferior lateral third of the clavicle. The medial segment attaches proximally at the superior surface of the acromion process. The posterior segment attaches proximally along the entire length of the inferior scapular border. All three segments attach distally to the deltoid tuberosity of the humerus (figure 13-19). To visualize the deltoid's segments, flex, abduct, and extend a straight arm against a resistance applied to the anterior, lateral, and posterior. It is normal for each of these segments to function together as a unit. However it is also normal for the segments to be differentially recruited during flexion, extension, and abduction of the humerus. For example the anterior deltoid is heavily recruited in the flexion of the humerus to the anterior along the sagittal plane, the medial segment contributes some to this movement, and posterior segment contributes even less to the movement. This little observation keeps body builders and general weight trainees using a large menu of isolation-type shoulder exercises hoping to develop the muscles individual segments maximally. As mentioned several times throughout this text (repetition is good for learning), isolation of small muscle masses is likely not the best approach to rapid and large-scale muscular development. Recruitment of individual muscles or segments of muscles does not produce a satisfactory anabolic stimulus to drive optimal fitness gain (strength or mass). It would be more beneficial to select an exercise that recruits the entire deltoid mass as equally as possible to drive uniform development. In this instance, we can abduct our humerus out to about 80 degrees in the coronal plane and alternatively flex and extend the shoulder joint and recruit the entire deltoid mass – do a press, an ancient and useful exercise. The deltoid is an important prime mover of the shoulder due to the wide variety of motion possible given its multiple attachments around the very mobile glenohumeral joint. It is also a very important stabilizer of the glenohumeral joint, contraction of the deltoid pulls the humerus and the glenoid fossa together, and does so more forcefully and effectively than any other shoulder muscle or group of muscles.

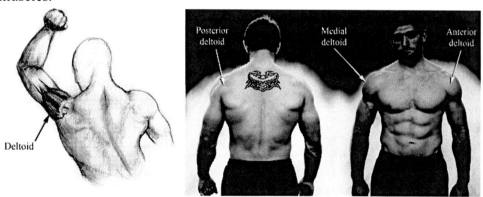

Figure 13-19. The deltoid.

Pectoralis major – The pectoralis major is likely one of the two most recognized muscles in the body. For good or ill a huge number of adolescent and adult males think that having big pectoralis major muscles is the ultimate signature of masculinity. It is an important muscle of the shoulder joint, but attitudes towards big "pecs" in men is a artificial social construct, barely 60 years old, driven by the evolution of comic book superheros and Hollywood leading men. All you need to do to verify this is to read, look at pictures, and watch movies dating back six or so decades.

The pectoralis major is a two segment muscle with the both segments sharing a common distal attachment to the lateral lip of the bicepital groove just inferior to the greater tuberosity (figure 13-20). The clavicular portion (pars clavicularis) attaches proximally to the inferior and medial half of the clavicle. This clavicular segment is superior and represents the smaller of the two segments. The sternal segment (pars sternalis) represents the bulk of the muscle, attaching proximally along the upper two thirds of the lateral border of the sternum. Both segments also have minor tendinous connections to the ribs near their major proximal attachments. Identification and palpation of the pectoralis major is easy. Flex the shoulder joint to the anterior at 90 degrees along the sagittal plane then adduct the humerus (the function of the pectoralis major) against resistance. An easy way to do this is to place the palms together, move the upper arm up parallel to the floor, and push towards the middle. Repeat this with the humerus at 10 degrees (forearms just in front of abdomen) and at 30 degrees (elbows at about the same level as the aereola). Note the difference in recruitment of muscle segments. At 10 degrees from humeral verticality, the sternal segment is most strongly active. At 90 degrees the clavicular segment is more dominant. At 30 degrees of humeral angle you will find relatively even recruitment. This simple observation should suggest that for optimal muscle recruitment, this is the angle of choice for use with exercises such as the bench press. This angle provides a favorable range of motion for maximal muscle movement and recruitment, this does not mean this is the best angle for powerlifting performance where moving as much weight as possible through the minimal legal distance is the goal.

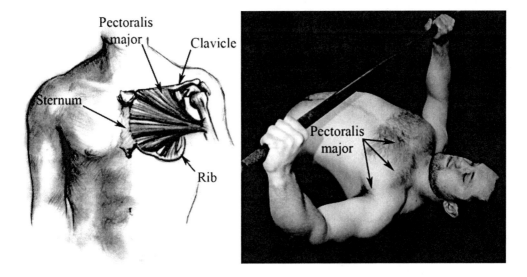

Figure 13-20. The pectoralis major.

Figure 13-21. Demonstration of the pectoralis major with an isometric exercise.

Biceps brachii – It will come as a surprise to many that the most recognizable "arm" muscles, the ones that flex the elbow, are actually shoulder muscles too. The biceps brachii do act distally to flex the elbow (considered in the following chapter), but they also act proximally relative to their frequently ignored attachments to the scapula (figure 13-22). The *short head* of the biceps brachii attaches proximally to the anterior point of the coracoid process and distally to the radial tuberosity (on the radius bone of the forearm), thus crossing two joints. This means that the short head of the biceps can, if the hand is held static, pull the scapula up and/or tilt it to the anterior. This attachment arrangement also produces glenohumeral flexion as a proximal function. This especially occurs in chin-ups, pull-ups, or during climbing. In a standing or seated position however, with a resistance in the hands or on the arms, the scapula will most likely be held static and the short head of the biceps brachii will carry out its distal function, flexing the elbow. Since so many trainees want to develop their biceps maximally, this structural and functional arrangement means that a correct bicep curl with a barbell or dumbbell should be comprised of first, flexion of the elbow followed by about 20 degrees of humeral flexion. In this manner both the biceps *long head* and short head are loaded effectively for better development throughout the muscle's complete range of motion (figure 13-23).

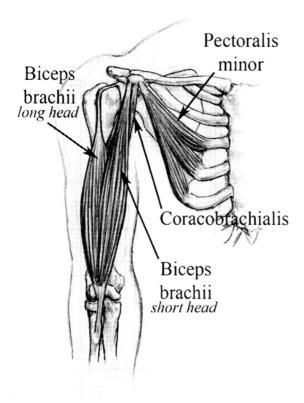

Biceps brachii *long head*

Pectoralis minor

Coracobrachialis

Biceps brachii *short head*

Figure 13-22. The biceps brachii, coracobrachialis, and the pectoralis minor.

It is interesting to note that biceps is the correct singular use of the Latin, not bicep (eg. "I have one massive bicep on my right arm" and "I have a total of two biceps"). The correct Latin plural would be " bicipites". The common misuse of bicep as singular and biceps as plural is so prevalent that most clinicians and practitioners would not easily recognize a reference to the bicipites brachii. As such the mal-use is begrudgingly acceptable modern parlance.

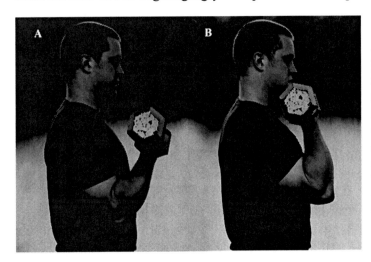

Figure 13-23. Most people think the bicep is maximally developed by simple elbow flexion - a curl (A). But in reality, complete development can only occur with a short extension of the humerus at the end of the motion, the biceps proximal function (B).

Coracobrachialis – The coracobrachialis is a small muscle that attaches proximally at the anterior point of the coracoid process (one of three to do so) and distally to the medial surface of the humerus – opposite to the deltoid tuberosity (figure 13-24). The origin of the name of the coracobracialis is derived from its attachments on the coracoid process (the arm) and on the brachium. It functions to flex and adduct the humerus. If someone is lean enough, placing ones arm parallel to the floor and attempting horizontal adduction should enable the muscle's small mass palpation on the inner aspect of the upper arm at the axilla (armpit) in between the biceps and triceps.

Pectoralis minor – The pectoralis minor attaches proximally to the upper and outer aspect of the 3rd, 4th, and 5th ribs. It lies underneath the pectoralis major and runs to the superior and medial side of the coracoid process of the scapula, its distal attachment (figure 13-22). Its deep position relative to the pectoralis major prevents its palpation. Contraction of the pectoralis minor pulls the scapula down and forward to a small degree (depresses the point of the shoulder and brings the scapula close to the thoracic ribs).

Latissimus dorsi – Although many people think of the latissimus dorsi as a very broad (latissimus means "broadest" in Latin), very large superficial muscle of the mid- to low back, the one giving the upper body its "V" shape, it is a very strong contributor to shoulder function. In fact, it carries out five movement roles at the shoulder, some more important than the others: extension, adduction, horizontal abduction, flexion from an extended position, and internal rotation. The muscle also contributes to extension and lateral flexion of the lumbar vertebrae. It is located lateral and inferior to the trapezius. It has multiple proximal attachments on the lower back including the lower four ribs (lateral and posterior aspect), the vertebral column from the sacrum up to the mid-thoracic level, a very small attachment at the inferior angle of the scapula, and on the posterior crest of the ilium (figure 13-24). The latissimus dorsi passes under the axilla (armpit) around the humerus to the anterior and distally attaches along the bicepital groove between the pectoralis major and teres major.

The mass of this muscle is indicative of its dominance over humeral position. It can pull the humerus to the posterior and internally rotate the humerus with great force. In concert with the deltoid, it is responsible for both gross movements AND joint stability during shoulder movements. You can visually identify the latissimus dorsi easily in any photograph of the back of a body builder. In normal people it is still easy to see and/or palpate in a seated individual. By leaning forward in the chair, palm on the shin just below the knee, pushing straight back against the shin with a straight will tense the muscle preferentially. The muscle will then be palpable from just below the axillary crease (bottom of armpit) down the anterior inferior aspect of the muscle and back to its attachment on the posterior iliac crest.

Triceps brachii – The triceps brachii is a three headed muscle, each head arising from different origins and joining together as the heads approach the elbow. The triceps suffer from the same linguistic singular-plural problem as the biceps. The long head of the triceps brachii attaches proximally to the scapula, along the lateral border immediately inferior to the glenoid fossa and distally to the ulna at the olecranon process (the point of the elbow) (figure 13-25). The long head of the triceps brachii can and does contribute to the muscle group's distal function of elbow extension. If the arm is held static, contraction of the long head can also pull the scapula down and lateral. However, its proximal attachment to the scapula gives it another overlooked proximal function, the scapula held stable and the humerus being adducted.

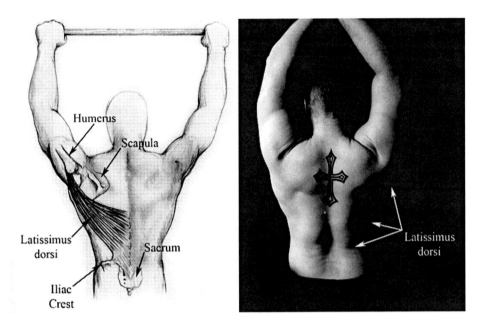

Figure 13-24. The latissimus dorsi is stretched when the arms are overhead.
While rows, pull-downs, and other isolation exercises are frequently suggested
as the best exercises for latissimus dorsi development, if morphology of
participants is any indication, deadlifts, cleans, and snatches may be superior.
Powerlifters and weightlifters develop very thick and strong latissimus dorsi
muscles as the pulling motions they routinely train strongly recruit the muscle's
adduction and internal rotation functions, and does it under larger loads than
possible with isolation exercises.

Serratus anterior –The serratus anterior has a characteristic saw-tooth
appearance with several visible and distinct segments attaching proximally to
the upper eight ribs and distally to the anterior medial surface of the scapula
along its inferior border (figure 13-26). The serratus rotates the inferior angle of
the scapula, protracts the scapular laterally around towards the front of rib cage,
and also isometrically holds the scapula in close proximity to the posterior
aspect of the ribs. Individuals who box or do lots of push-ups will display well
developed serrati muscles. Palpation is fairly easy. With the arm abducted or
flexed overhead, providing a resistance at the elbow (stand in front and pushing
gently backwards) will cause a serratus anterior contraction. The muscle then
can be palpated with the flat of the four fingers down along the mid-axillary line
(arm-pit and down). During most contractions of the serratus anterior, the ribs
will be more stable than the scapula (less potential for mobility). As such, the
scapula will move along the line of action dictated by the orientation of the
muscle fibers. An exception to this occurs with any exercise or sporting activity
involving suspension of the body by the hands and arms – chins, rope climbing,

205

certain gymnastic movements, for example. In suspension, the scapula will be the most stable structure in the system and the serratus anterior muscles will be used to assist in elevating the thoracic segment of the body towards the scapula/arm/hand/apparatus system.

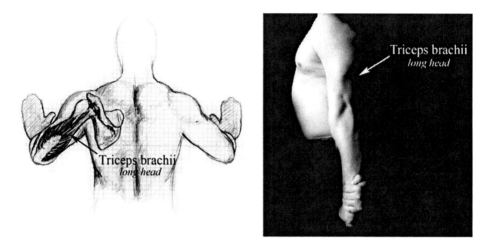

Figure 13-25. The triceps brachii long head is active at the shoulder.

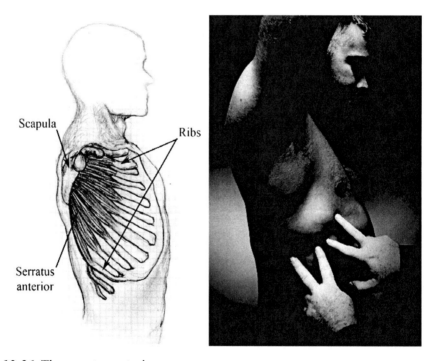

Figure 13-26. The serratus anterior.

Rhomboideus – The rhomboideus has two sections (*major* and *minor*) that appear, as the rhombus prefix would suggest, a diamond shape. The minor is smaller and superior to the major. Although there are two distinct muscles, they function as a single unit. Located beneath the lower trapezius segments and non-palpable, the rhomboideus minor muscle attaches proximally to the spinous processes of the seventh cervical vertebra along with the first and second thoracic vertebrae. The rhomboideus major attaches proximally to the spinous processes of the second, third, fourth, and fifth thoracic vertebrae. Both the major and minor attach distally to the medial border of the scapula at about the level of the scapular spine, with the major's attachment extending well down to the scapula's inferior angle (figure 13-27). Contraction of the rhomboideus will generally cause retraction of the scapulae (pulling them closer to the midline of the body - or squeeze the shoulder blades together). The rhomboid major helps to hold the scapula (and thus the upper limb) onto the ribcage. It also acts to downwardly rotate the scapula, with respect to the glenohumeral joint and works synergistically with the levator scapulae to elevate the medial border of the scapula.

Levator scapula – The levator scapula is a muscle found deep and inferior to the trapezius (figure 13-27). It attaches proximally to the transverse processes of the atlas, axis, third and fourth cervical vertebrae. Distally it attaches to the superior medial scapular border. This is a load bearing muscle that elevates the medial scapula and as such it will fatigue during activities like backpacking or carrying heavy loads as in the Farmer's Walk. It is not palpable due to the overlying trapezius.

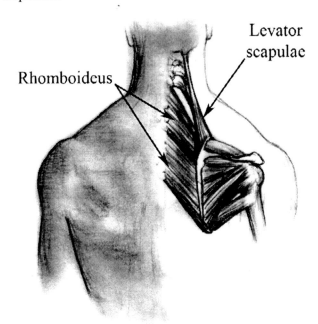

Levator scapulae

Rhomboideus

Figure 13-27. The rhomboideus.

Supraspinatus – The name provides a clue to the muscle's location, above a spine (figure 13-28). It lies above the scapular spine with its proximal attachment along the curvature formed by the spine and the body of the scapula superior to the spine. This depression on the body of the scapula is called the supraspinous fossa. The muscle attaches distally to the greater tuberosity of the humerus at its most superior point. This superior attachment means it is able to generate very little torque around the joint. The distal tendon passes beneath the acromion and superior to the glenohumeral joint. Its position relative to the joint makes it an abductor, but its small size limits its contribution to movement in only about the first 15 degrees of abduction. After that point, the co-contracting deltoideus assumes even more of the original abduction load and the supraspinatus is relegated to an essentially isometric stabilization role. The supraspinatus is covered by the overlying superior segments of the trapezius and is thus too deep to be palpated.

Infraspinatus – The name infraspinatus gives us another excellent idea of this muscle's location. It is inferior to the scapular spine, along the infraspinous fossa. The spine of the scapula separates the supraspinous and infraspinous fossae. It attaches distally to the greater tuberosity of the humerus (figure 13-28). The infraspinatus is an external rotator of the glenohumeral joint and assists in abduction of the arm. However, as with the supraspinatus, its relatively small size and its very short lever arm on the humerus means that the muscle is capable of producing very little torque around the joint and will act primarily as a stabilizer.

Teres major – The teres major is a fairly small muscle that lies superior and approximately parallel to the latissimus dorsi, lateral to trapezius and inferior to the posterior deltoid (figure 13-28). It attaches proximally near the posterior inferior angle of the scapula and runs under the axilla and attaches distally on the anterior surface of the humerus, at the bicepital groove medial to the distal attachment of the latissimus dorsi. The teres major carries out similar functions to the latissimus; a medial humeral rotator, an adductor of the humerus, and a synergist with the latissimus dorsi in pulling a previously elevated humerus down and back (active in the pullover barbell exercise). Given its small size and poor leverage position, it is not a large contributor to forceful humeral movement, rather it acts with the latissimus dorsi to stabilize the humeral head in the glenoid fossa. The teres major can be visualized on lean individuals using the same approach as in identifying the latissimus dorsi, with the exception that the muscle is traced back to the scapula, not the iliac crest. If more subcutaneous fat tissue is present palpation is the only means of identification.

Teres minor – Again, the name should indicate to us that this muscle is not too far from teres major. It is, in fact, located just superior to the teres major. The teres minor attaches proximally from just below the glenoid fossa then downwards about two thirds of the way along the lateral border of the scapula. (figure 13-28). As the muscle approaches the axilla, it passes posterior to the shoulder joint and attaches distally to the greater tuberosity, just lower than the attachments of the supraspinatus and infraspinatus. The teres minor functions to pull the humeral head into the glenoid fossa and to externally rotate the head of the humerus to the posterior. Both the teres minor and infraspinatus may be palpable on a well developed or lean individual. With strong horizontal abduction and outward rotation of the humerus, you can generally find these muscles in relation to a triangle formed by the inferior line of the posterior deltoid, the lateral border of the trapezius, and a line formed by the superior aspects of the teres major and latissimus dorsi.

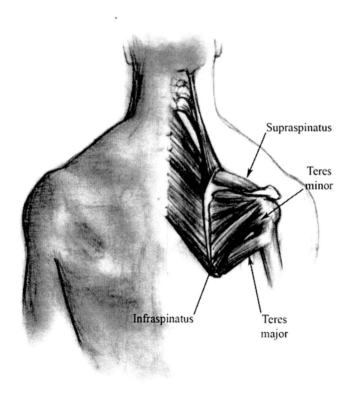

Supraspinatus

Teres minor

Infraspinatus

Teres major

Figure 13-28. The supraspinatus, infraspinatus, teres major and minor.

Subscapularis – Sometimes this muscle is confused with the infraspinatus, but breaking down the name into its roots makes the location of the muscle easily distinguishable. The prefix "sub" means lying below or under. "Scapularis" refers obviously to the scapula so the subscapularis lies under the scapula in the space between it and the underlying rib-cage. This location and the fact that its distal elements are under the deltoid means that the muscle is not palpable. The subscapularis attaches proximally across the majority of the anterior and inferior surface of the scapula, the medial and lower two thirds of the scapula's anterior surface. It is a broad and roughly triangular shaped attachment that thins as it approaches the axilla. The muscle attaches distally on the lesser tuberosity of the humerus (figure 13-29). This anterior attachment results in internal rotation of the humeral head during contraction. This is a moderately large muscle that can produce a significant amount of tension drawing the head of the humerus into the glenoid fossa, thus serves as a stabilizer of the shoulder and defense against dislocation.

Anterior view

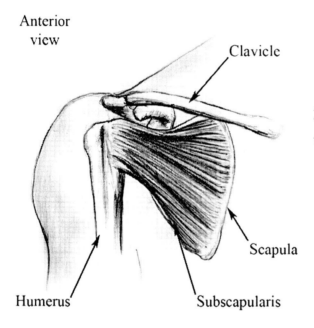

Figure 13-29. The subscapularis. The rib cage would be in front of the scapula here.

MOVEMENTS OF THE SHOULDER

There are a number of movements of which the scapulae are capable. The list below is illustrative but remember that the scapular and move up, down, side-to-side, rotate left, rotate right, tilt forward, tilt back, tilt to the side, and any combination in between (figure 13-30, 13-31).

Adduction of the scapula is their movement towards the vertebral column. Sometimes called retraction. A layman's description might be squeezing the shoulder blades together.

Abduction is the movement of the scapula away from the vertebrae. A layman's description might be rolling or rounding your shoulders forward.

Elevation is the drawing of the scapulae in a superior, upwards direction. The common description of this is a shrug of the shoulders.

Depression is the pulling of the scapulae in an inferior, downwards direction. This can be accomplished by getting someone to lift their chest and pull their shoulders straight down.

Upward rotation is usually defined as a movement around a sagittal axis that passes through the center of the scapula, however, in actuality scapular movement does not occur around an axis at the geographic center of the scapula, it rotates around a point just inferior to the scapular spine and a few centimeters medial to the middle of the complete length of the spine. Upward rotation occurs when the inferior angle of the scapula moves to point more laterally, usually accompanying humeral abduction.

Downward rotation is again usually defined as a movement of the scapula and inferior angle of the scapula around a central axis towards the anatomical midline.

The movements of which the glenohumeral joint is capable of abduction, adduction, flexion, extension, and rotation. These movements can occur in any plane within which the shoulder joint is capable of movement. They can occur singly or in combination. If the arm is raised directly in front of the body (from anatomical position), the shoulder is flexed. If the arm is raised to the rear, it is extended. Raised laterally to the side, it is abducted. Raised across the front of the body, it is adducted. If one starts at anatomical position and raises the arm forward, at the same time moves the thumb to point at the floor, while also moving the angle of the humerus to parallel to the frontal plane, simultaneous flexion, abduction, and rotation has occurred at the shoulder joint assembly. Multi-axial joint movements can be very complex and require a great deal of neural synchrony and coordination, this is another reason that exercises that involve more musculature and that are not artificially restricted (restricted as in exercise machines) are the most effective and beneficial ways to exercise.

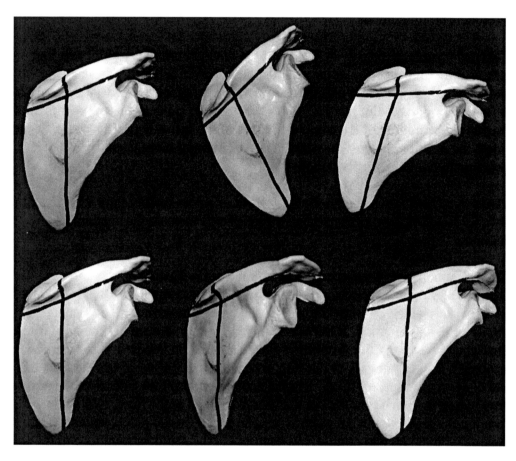

Figure 13-30. The axes of movement of the scapula. Movement of the scapula occurs around a common point, where the x, y, and z axes intersect. In the images to the left the x and y axes are indicated by the lines lying parallel to the scapular spine and vertically, one third the total width of the scapula in from the medial border. The y axis intersects, anterior to posterior, at the intersection of the x and y. The rotation of the scapula during movement occurs around this central point as indicated in the photographs. Top – rest, medial rotation, lateral rotation. Bottom – rest, posterior rotation, anterior rotation.

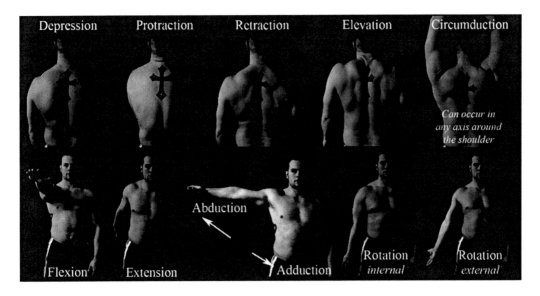

Figure 13-30. The movements of the shoulder.

OFF THE CUFF COMMENTARY

Note that the arm is suspended relative to the axial skeleton with very little bony support, rather it is muscle and ligament that attach it to the thoracic architecture. This arrangement allows a great deal of mobility but also provides us with a problem with joint stability in high velocity or high force conditions. A popular topic in the exercise and sport professions relative to this is the "Rotator Cuff".

The supraspinatus, infraspinatus, teres minor, and subscapularis form a functional group of muscles commonly known as the rotator cuff (SITS is an often used acronym for remembering the muscles included). This group of small muscles contributes somewhat to motion and more so to stability of the glenohumeral joint. The manner in which the tendons are positioned around the joint provides the stability native to this combined structure (figure 13-32). The tendons of the infraspinatus and teres minor are posterior to the glenohumeral joint and attach to the humerus. The tendon of supraspinatus passes directly superior the joint and also attaches to the humerus. The tendon of the subscapularis lies anterior to the glenohumeral joint and also attaches on the humerus. Note that the attachments of each of these muscles are very close to the end of the humerus. Now think of the basic rules of a lever. It should be apparent that these muscles' primary role is not gross movement of the shoulder, rather their primary purpose in all shoulder movement is to keep the humeral head firmly seated in the glenoid fossa.

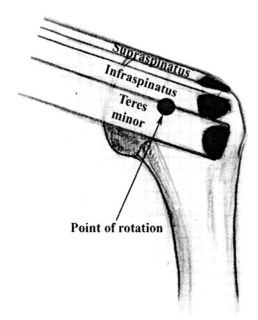

Point of rotation

Figure 13-32. The rotator cuff muscles attach distally to the femoral head and neck. The proximity of attachment to the rotational center of movement makes for a less than efficient motor system. However, the line of action of the rotator cuff muscles does prove to be an effective orientation to resist separation of the humeral head from the glenoid fossa.

It is a common misconception that if the rotator cuff is injured, such as a muscle strain or sprain, that rehabilitation should be carried out using extremely little resistance and primarily with internal and external rotation exercises (figure 13-33). This type of exercise IS adequate for initial use in re-establishing range of motion, but the isolation of individual small muscles does not improve the strength and integrity of the shoulder joint to the point it can resist repeated stress or further injurious insult (what caused the problem to start with). The fact of the matter is, that the injury to the rotator cuff was likely due to the overall musculature of the shoulder not being strong enough to resist the force that caused the injury - a pitcher throwing a fastball or an outside hitter repeatedly producing large amounts of tension along the length of the joint, proximal to distal, through the simple act of forceful throwing or hitting are the most common sports causes. A simplistic analysis is that they are producing a force that is attempting to displace the humerus from its seat in the shoulder joint (throw the arm out of socket). If the total musculature surround the joint is too weak to withstand the tension forces, the weakest links, the small rotator cuff muscles can be torn to some degree.

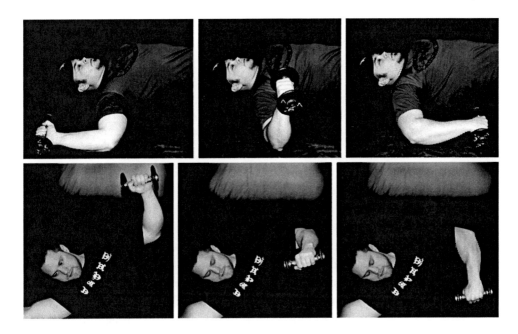

Figure 13-33. Typical rotator cuff rehabilitation exercise.

Doing lots of internal and external rotation repetitions with 3 pounds cannot induce a strength and rehabilitation stimulus. Think of it like this, if I can do 50 push ups and bench press 185 pounds, if I get my push ups up to 100 it does not mean I can bench press 370 pounds. No. I will just be able to do 100 push ups and my bench press will not increase by leaps and bounds, if at all. Lots of repetitions in internal and external rotation exercises develop local muscular endurance in the rotator cuff not the strength and ability to keep the humeral head seated.

Prevention and rehabilitation of injuries to the rotator cuff should be comprised of progressively weighted exercises carried out in multiple axes and directions that recruit as much of the shoulder musculature as possible, not miniature weights or bands in restricted ranges of motion. Presses and dips along the vertical planes and bench presses and rows along the transverse planes can strongly contribute to developing dynamic strength of the primary movers of the shoulder while at the same time strongly driving the development of dynamic and isometric strength of the rotator cuff. These exercises thus become preventive and rehabilitative when used progressively.

Somewhat related to this is the issue of shoulder impingement. "Impingement" refers to the physical compression of soft tissues between head of the humerus and the acromion process (figure 13-34B). Inflammation of the tissues lying in this space may cause shoulder pain and is frequently referred to as shoulder impingement syndrome, as the inflamed and enlarged tissues are impinged upon by skeletal elements during movements of the shoulder. Other than that inflammation is present, the underlying cause of shoulder impingement syndrome is not well understood. Improper postural mechanics, anterior-posterior muscle strength imbalances, poor shoulder flexibility, and the physical shape of the acromion process (one of the four identified acromial shapes is associated with a higher rate of shoulder impingement syndrome) have all been proposed as causes. It is important to note here, and contrary to common beliefs, clinical and laboratory investigations have not indicted overhead exercises as a cause of shoulder impingement syndrome. Regardless, many less informed coaches and clinicians have adopted the unsupported position that overhead movements are contraindicated in physical conditioning programs. It is likely that this bias is based upon the higher than normal frequency of shoulder impingement syndrome in "overhead" athletes, those that throw or strike repeatedly in training and competition. This is an inappropriate extension of an observation. These athletes are involved in activities that produce tension across the shoulder joint. Normal work and conditioning activities, most common training exercises, produce compression forces across the joint so any conclusions drawn from those injuries from tension-producing activities are in err.

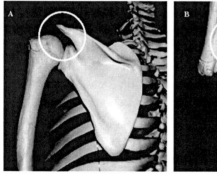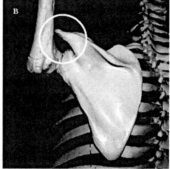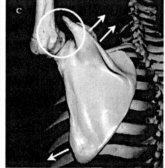

Figure 13-34. Normal acromiohumeral space (A). Impingement with abduction (B). Correction of impingement with simple trapezius and serratus anterior contraction (C).

Three of the four suspected causes of shoulder impingement syndrome can be prevented or treated with weighted conditioning activities through a complete range of motion. Strengthening the musculature associated with shoulder posture, strengthening the posterior shoulder compartment, and establishing the ability to move the shoulder joint through a complete range of motion are all valid and important goals in the prevention and treatment of shoulder

216

impingement syndrome. When performed correctly by healthy individuals, conditioning activities, will assist in prevention of shoulder problems not produce them.

Despite the logic of performing exercises to strengthen the shoulder through its complete range of motion and also improve its contractile endurance, there will continue to be detractors to overhead exercises for conditioning. A simple anatomical examination of the any lift so questioned, done correctly, can provide a persuasive argument for its safety.

In regards to shoulder impingement, when we consider the effect of overhead movement of the arm on the space between the humeral head and acromion, we cannot simply consider the rotation of the humerus in the glenoid fossa. If we do, we will arrive at a condition where the acromion process will contact the humerus and any tissues between the two will be impinged. This one-dimensional analysis does not consider the multiple structural features and movements that accompany overhead movement of the arm. The movement of the scapula must also be considered in any analysis. In overhead pressing movements, the orientation of the scapular spine and its lateral terminus, the acromion process changes by rotation on its axis and becoming more vertical. Competent exercise professionals teach an active engagement of the trapezius in order to more fully develop this important shoulder muscle. A vital and additional benefit of this practice is that the acromion process is held at a distance from the humerus by virtue of the trapezius "shrug" (figure 13-34C). This results in an elimination of the basis for conventional arguments against overhead movements or in performing them for many repetitions as impingement is unlikely.

There is virtually no place in conditioning where the tenets of clinical rehabilitation can or should be applied to able-bodied trainees. To do so, train the healthy like the injured or diseased, would compromise their ability to perform life activities and exercises effectively and would further predispose them to injury during their execution. Progressive, full range of motion conditioning activities, that reflect the demands of sport, work, and daily life is appropriate exercise prescription. In the case of shoulder impingement, full range of motion overhead lifting is appropriate, performance enhancing, safe, and will prevent the very injuries it is proposed to cause.

"To say that a man is made up of certain chemical elements is a satisfactory description only for those who intend to use him as a fertilizer."

*- **Hermann Joseph Muller***

14 - THE ELBOW

The elbow is a generally considered a simple hinge joint formed by the distal end of the humerus along with the ulna and radius of the lower (fore) arm. In reality it is a collection of three joints; the humeroulnar, the humeroradial, and the proximal radioulnar. The humeroulnar joint, the articulation of the distal humerus and proximal ulna, and the humeroradial joint, the articulation of the distal humerus and the proximal radius conform to the hinge joint classification. The basic movements of the elbow are flexion and extension, making activities such as pulling objects toward us or push things away from us possible. Although the elbow's movements are quite simple, the range of motion provided by combining the possible actions of the shoulder and elbow joints as a system is quite complex, providing for precise positioning of the arm in a large space surrounding the body. The proximal radioulnar joint, the side-by-side articulation of the proximal radius and ulna, is a pivot joint that comes more into play in the actions of the hand.

The architecture of the two humerus-associated joints makes the elbow a very stable joint system with few exercise related pathologies. The primary problems associated with sport and the elbow are:

> *Tendonitis* - An over-use injury as in tennis elbow, golfer's elbow, archers elbow, and others. More precisely called epicondylitis and can occur as chronic pain and inflammation on either the lateral (as in tennis) or medial side (as in golf).

> *Dislocation* - The disarticulation of one or more of the bones of the elbow. Only about 6 persons per 100,000 will get an elbow dislocation each year.

> *Fracture* - A break in on of the bones of the joint at a point proximal to the joint. Generally occurring as a result of a fall or other forceful contact.

BONES

Each of the three bones of the elbow possess anatomical features that contribute to the function of the joint. The characteristics of these functional bumps, ridges, and knobs, provide for the stability inherent in the joint and the musculotendinous attachments required to make it mobile (figure 14-1).

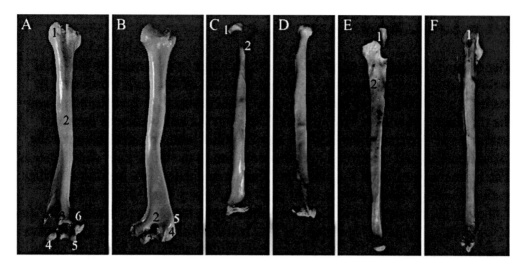

Figure 14-1. Anterior and posterior views of the three bones of the elbow and their features. (A) Humerus, anterior view. 1-head, 2-shaft, 3-condyles, 4-capitulum, 5-trochlea, 6-supracondylar ridge. (B) Humerus, posterior view. 1-Medial epicondyle, 2-olecranon fossa, 3-trochlea, 4-lateral epicondyle, 5-supracondylar ridge. (C) Radius, anterior view. 1-head, 2-radial tuberosity. 3-styloid process. (D) Radius, posterior view. (E) Ulna, anterior view. 1-trochlear notch, 2-ulnar tuberosity. (F) Ulna, posterior view. 1-Olecranon, 2-styloid process.

Humerus - The distal end of the humerus has a unique set of bony features providing for articulation with the radius and ulna.

The trochlea is the roughly hour-glass shaped feature on the distal end of the humerus and articulates with the trochlear notch of the ulna. The head of the radius articulates with the capitulum of the humerus, the ball shaped feature just lateral to the trochlea. The trochlea and capitulum are the rounded, smooth, knuckle-like surfaces at the anterior and distal end of the humerus. It is around these structures that the elbow flexes and extends.

There are two outcroppings of bone flanking the trochlea and the capitulum. The medial epicondyle and lateral epicondyle (of the humerus) are just proximal, lateral, and superior to the trochlea and capitulum. It may seem odd to have epicondyles present when there are no structures called condyles observed, as was seen at the knee. But condyles really are present here. Condyles are knuckles or rounded projections and the trochlea and capitulum are condyles, they just have names. The epicondyles can be found by palpating the widest points of the distal humerus. The thumb and forefinger can generally span the elbow joint at its widest point from the posterior (figure 14-2). The epicondyles should be evident as two superficial bony bumps extending from the distal and widest point and upwards as they narrow into the shaft of the bone. The narrowing represents the supracondylar ridges, extending from the epicondyle upward along the lateral and medial borders of the humerus. The medial epicondyle will be larger than the lateral.

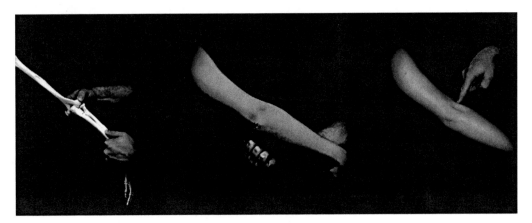

Figure 14-2. Locating the epicondyles of the humerus.

On the flip side of the distal humerus (posterior) there is a fairly significant depression called the olecranon fossa. The anterior end of the ulna fits snugly into this depression when the arm is straight (elbow extended). This provides a great deal of stability to the extended joint and a means of protection from hyperextension. It also creates a very hard physical limit to the range of motion of the elbow in extension.

Ulna - The ulna is a long prismatic shaped bone found on the medial side of the forearm (pinky side). The proximal end of the bone is the thickest, narrowing to a near point as it approaches the wrist.

The proximal end of the ulna features the prominent bony process, the olecranon process, or point of the elbow. It can be palpated in its position between and inferior to the epicondyles, most easily when the elbow is flexed (bent) (figure 14-3). On the anterior side of the ulna, the trochlear notch is found. The trochlea of the humerus fits nicely into this fairly deep "c" shaped notch. Just anterior to the notch is the coranoid process. The depth and width of these latter two feature provides additional support to the elbow's architecture.

Radius - The third bone of the system is the radius, the smallest of the elbow (figure 14-4). The radius runs from the elbow down to the wrist on the thumb side. It is a long bone, curved along its length, and crudely prismatic in shape.

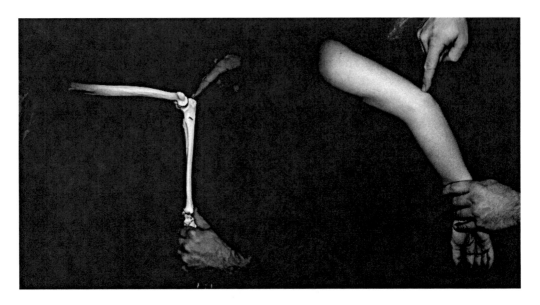

Figure 14-3. Finding the olecranon process, the distal end of the humerus.

The radial head, at the proximal end of the radius, articulates with the capitulum of the distal humerus and is just inferior to the lateral epicondyle. The ball shape of the capitulum and the nail-head looking radial head comprise a nicely designed pivot joint - the head of the radius rotates on the ball of the capitulum. Careful palpation just medial and inferior to the lateral epicondyle followed by rotation of the hand should reveal a bump that appears with inward rotation and disappears with outward rotation, the radial head.

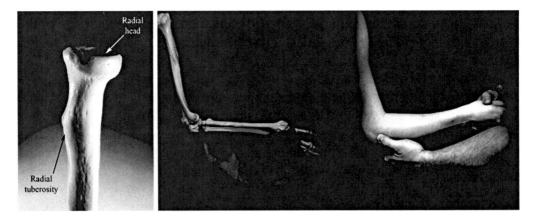

Figure 14-4. Features of the proximal radius (left). Finding the radial head (right).

The most important bony feature relevant to the elbow is the radial tuberosity. A number of muscles of the elbow attach to it.

JOINTS

Humeroulnar - Connecting the trochlea of the humerus to the the trochlear notch of the ulna, this is a hinge joint. There is an ulnar collateral (co-paired and lateral to the side) ligament present (figure 14-5). The anterior ulnar collateral ligament connects the medial epicondyle and the medial aspect of the coranoid process. The posterior ulnar collateral ligament attaches to the lower and posterior portion of the medial epicondyle and to the medial aspect of the olecranon. Both ligaments may start medially but they spread laterally and meet the criterion expected in their names. This particular ligament is vulnerable to injury in high velocity throwing athletes. Rupture of this ligament increases with the number of throws completed in a season of play. As few as 200 high velocity, high force throws over a period of months increases the risk of injury to this ligament by 63%. Surgical repair of the injury, Tommy Johns Surgery - after the first athlete to receive it, is quite robust and full recovery of prior function is expected. Johns went on to play professionally (baseball) for fifteen years after repair.

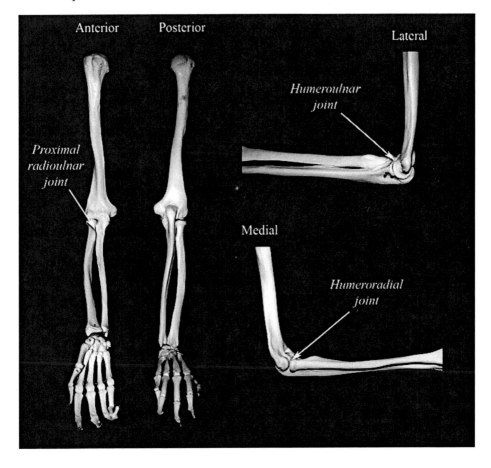

Figure 14-5. The joints of the elbow.

Humeroradial - This joint is the articulation of the capitulum of the the humerus and the radial head. Like the humeroulnar it is a hinge joint. The major ligament of this joint is the annular ligament that binds the radius to the ulna.

Proximal Radioulnar - "Proximal" is used here as there is a second radioulnar joint at the distal end of these bones. At the proximal end the articulation is between the radial notch of the humerus and the head of the radius. The ligament contributing to the stability of the proximal radioulnar joint, and to the humeroradial joint, is also the annular ligament. This ligament crosses and encircles the radial head and keeps it close to the proximal ulna. This is an important ligament, preventing radial dislocation. In fact, if this ligament was not present the biceps brachii would dislocate the radius with every contraction.

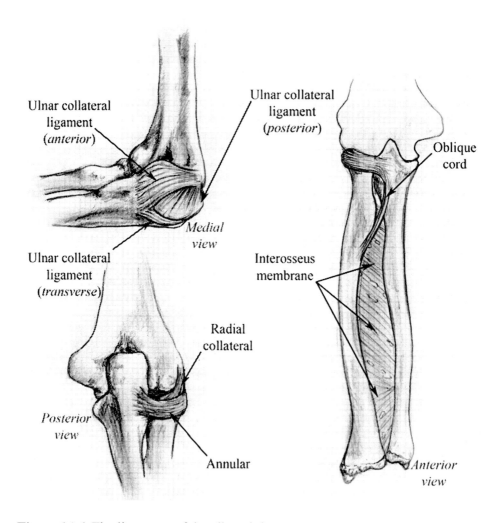

Figure 14-6. The ligaments of the elbow joints.

There is another ligament connecting the ulna and radius near their proximal ends, the oblique cord. The cord angles across the space between the proximal ulna at the base of the coranoid process to the proximal radius just inferior the radial tuberosity. The ligament helps prevent the separation of the two bones, limits rotation, and may actually reinforce bone structure and help prevent bending.

Another component of the multiple connections between the radius and ulna is the interosseous membrane. This is a sheet of fibrous connective tissue that runs diagonally between the radius and ulna. Much like the oblique cord, it maintains the integrity of the proximal radioulnar joint (it serves the same purpose for the distal radioulnar joint). It functions to prevent separation of the two bones, redistributes forces applied to the radius to the ulna, and resists asymmetrical longitudinal sliding along the longitudinal axis of the two bones.

MUSCLES

Lots of attention is paid to the muscles of the arm. But everyone seems to think that there are only two and that the primary work of the "bi's and tri's" is to bend and extend the elbow. Even though the elbow is a simple structure and the musculature here is confined to a small number of muscles, its not all about having exceptional double front bicep and side tricep poses for the beach. The elbow musculature is an integral part of an upper axillary system that is capable of both extremely refined and very powerful movement.

Biceps brachii - The first muscle considered is the biceps brachii, arguably the most exercised muscle group in the weight room. Virtually every muscle and fitness magazine on the newsstand includes curls as a strength-training program mainstay. The popularity of curls has its roots in the evolution of the bodybuilding counter-culture that began in the 1940's and that continues to evolve. This attention is likely unwarranted, as their function in this respect is to simply flex the elbow joint. There is nothing magic or of functional importance gained by having 20 inch "pythons". It is nice having strong biceps but they are small potatoes in the scope of power (and endurance) performance.

The biceps brachii attach proximally on the coracoid process (long head) and near the glenoid fossa (short head) then attach distally at the radial tuberosity (figure 14-7). This attachment to the radial tuberosity is of interest functionally. The proximal function of the biceps brachii has previously been covered. If the shoulder is held stable, a distal function is then revealed, simple flexion at the two humeral articulations of the elbow. What is frequently not considered is that the biceps brachii, by virtue of the attachment to the radial tuberosity, has an even more distal function, internal rotation (pronation) of the hand. This is accomplished by pulling the radius up and over the ulna at the radioulnar joint.

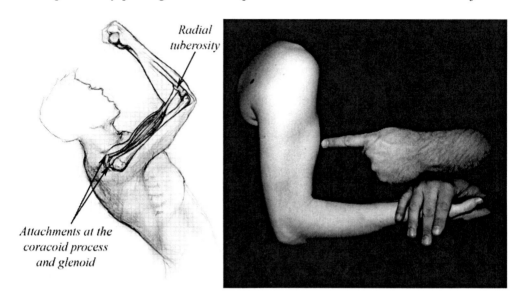

Radial tuberosity

Attachments at the coracoid process and glenoid

Figure 14-7. Locating and identifying the biceps brachii.

Visualization and palpation of the biceps brachii is easy. Flex the elbow isometrically against manual resistance. Palpate for the emergence of the tendon from the radial tuberosity and follow the muscle to its proximal attachment of the long head at the coracoid process. Depending on the level of muscularity and tone in the pectorals, this proximal tendon may or may not be palpable. The proximal attachment of the short head at the glenoid fossa is under the deltoid and cannot be palpated.

A fun thing to do is to watch the shape change in the biceps brachii muscle belly induced by supination and pronation. Observe the length of the biceps muscle belly while rotating the palm down and up (keeping the elbow bent at 90 degrees). This change in length is a function of the biceps' attachment to the radius and its rotation over the ulna. This simple anatomical observation has relevance to that dastardly exercise, the curl. If you are interested in exercising through the complete range of motion for maximal functional development, they should be done in supination (with palms up) and finish with about 10-20 degrees of humeral flexion (as described in the previous chapter). This provides resistance and a developmental stimulus through the longest range of motion possible of the muscle.

Brachioradialis - The brachioradialis has a proximal attachment to the lateral epicondyle and supracondylar ridge of the humerus and a distal attachment to the radial styloid process (the point of the distal end of the radius) (figure 14-8). The muscle flexes the elbow however it is small and is not a strong force producer. Rather it is its roles as a pronator or supinator of the wrist and hand - it can do either depending on the rotational position of the hand - that are its major function. Palpation of the brachioradialis is elementary, as it is quite superficial and readily visible and palpable on the thumb side of the forearm. Pronating the hand until it is in a "thumbs up" position than flexion of the elbow up against a resistance will make the muscle palpable. The muscle would be a contributor to elbow flexion while the palms face each other such as in lifting a pot, or if considered singly when lifting a mug.

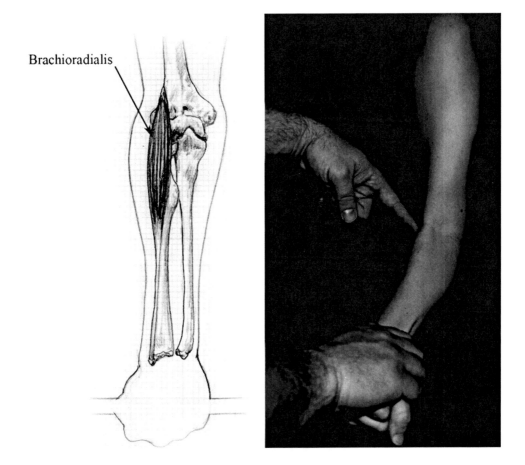

Figure 14-8. The brachioradialis.

Brachialis - Another flexor of the elbow, the brachialis (brachialis anticus) stretches from the anterior lower half of the shaft of the humerus down to the proximal end of the anterior ulna and its coronoid process (figure 14-9). The brachialis, like the biceps brachii, is a strong flexor of the elbow however given its attachment sites, closer to the elbow, it is at a mechanical deficit compared to the biceps and generates less of a motive force in flexion. This muscle does lie beneath the biceps brachii but its medial aspect is palpable. Supination of the forearm and flexion of the arm against resistance will make the muscle belly rigid when the fingers are placed on both sides of the biceps tendon.

Supinator - The supinator is a deep muscle of the elbow, attaching proximally to lateral epicondyle of the humerus and distally to the lateral aspect of the radial tuberosity and along the body of the bone a short distance (figure 14-8). Although it crosses the elbow, its size and orientation provides little ability to produce flexion, rather it acts as its name implies, as a supinator rotating the radius laterally.

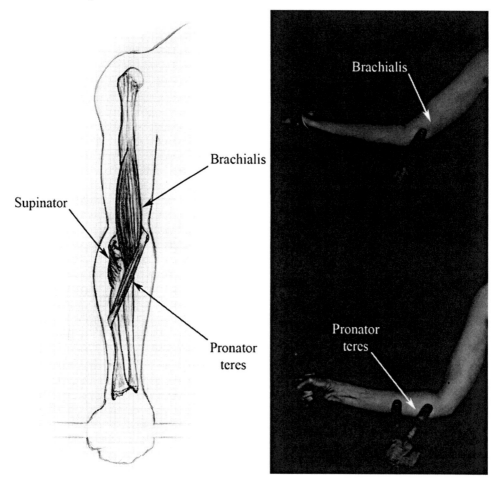

Figure 14-9. Identifying the brachialis, pronator teres, and the supinator.

Pronator teres - The pronator teres attaches proximally to the medial aspect of the supracondylar ridge of the humerus and distally to the proximal end of the radius, to about half way down its length (figure 14-8). It crosses the elbow joint and can be considered an elbow flexor but since it attaches medially to the medial epicondyle and laterally to the radial tuberosity, it is also a potent pronator of the hand. Palpation of the muscle is done through repeated pronation of the hand against resistance followed by unresisted supination. The pronator teres and the brachioradialis flank the biceps brachii tendon, medially and laterally respectively. The slight skin depression within the "V" formed by the two muscles on the upper forearm is called the cubital fossa - the area from which you usually get blood drawn.

Triceps brachii - The triceps brachii as its name suggests is morphologically, a muscle having three segments of the arm. There are three fascicles (bundles) of muscle making up the muscle body - the long, medial, and lateral head - each with a separate proximal attachment and a shared distal attachment (figure 14-9). The long head, aptly named as it attaches proximally to the scapula, inferior to the glenoid fossa (as discussed in the shoulder chapter), attaches distally via a shared triceps tendon to the olecranon of the ulna. The lateral head attaches proximally to the posterior humerus from an area just inferior to the greater tuberosity and down about a third the length of the shaft. The lateral and long heads are the most visible and palpable portions of the muscle. The medial head lies mostly under the lateral head, attaching proximally to the humerus along a narrow triangle stretching along the lower three fourths of the humerus. All three sections attach distally to the olecranon process of the ulna via a fusion into an elongated and central tendon.

The long head has two functions as it crosses both the shoulder and elbow. The first being its proximal function at the shoulder, the second and common to all three heads of the triceps brachii is extension of the elbow joint.

Anconeus - The anconeus is a small, roughly triangular muscle attaching proximally to the point of lateral epicondyle of the humerus and distally to the olecranon process and downwards a short distance (figure 14-9). Its small size and unfavorable orientation to produce torque suggests that it is not a strong elbow extensor although it does assist. It can also function as a supinator and as a tension provider to keep the joint capsule taught thus preventing impingement. The anconeus may be palpable in the area between the lateral epicondyle and olecranon process with mild flexion of a hanging arm then gentle extension.

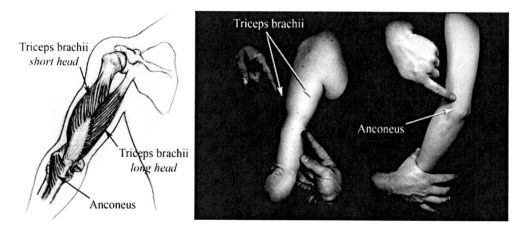

Figure 14-9. Locating the triceps brachii and the anconeus.

MOVEMENTS

The humeroulnar and humeroradial joints are hinge joints and are capable of flexion and extension (figure 14-10). Flexion involves the movement of the hand and forearm towards the shoulder via rotation around the joint. Extension at the humeroulnar joint is the opposite of flexion and is the movement of the hand and forearm away from the shoulder. The radioulnar joint rotates in two directions thus facilitating pronation and supination (figure 14-11). Pronation is the inward rotation of the forearm and hand, putting the hand palm down. Supination is the outward rotation of the forearm and hand, putting the palm up.

In respect to whole body locomotion, the structure of the elbow generally reflects a sex-related difference. When the elbow is extended, the line of humeral identity is not shared with the radius an ulna. In anatomical position, the angle of the lower arm to the upper arm deflects away from the body, this is called the carry angle (figure 14-12). There is quite a range of normal angles, from as little as 3 degrees up to as many as 29 degrees. Women generally have a more acute (higher number) carry angle than men, a putative result of human development where the wider hips of women require such an angle to enable impediment free arm swing during walking. An extreme angle away from normal is termed cubitus valgus. An angle toward the body is called cubitis varus. Both, especially in severe cases, carry with them possible movement limitation and risk of injury in sport and exercise, but not to the extent to prevent participation in either.

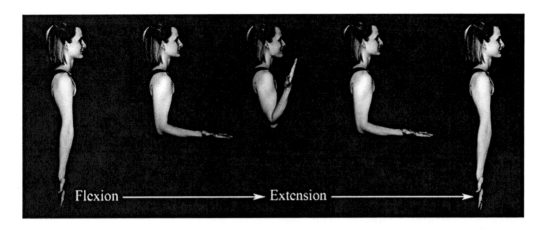

Figure 14-10. Flexion and extension of the elbow.

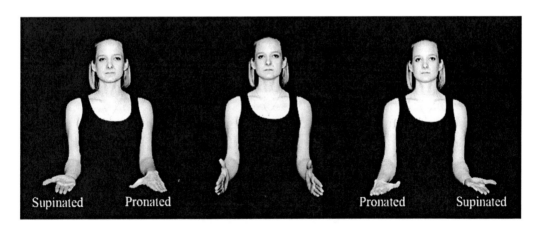

Figure 14-11. Rotation of the radioulnar joint allows supination and pronation.

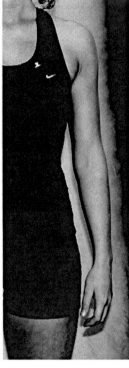

Figure 14-12. Relaxed carry angle of a male and female.

The carry angle represents complete extension of the elbow. Complete extension of the elbow is an important technical issue during pulling objects from the floor and in holding objects over the head. In the former case, a completely extended arm represents a condition where maximal force can be transmitted from the body, what is creating the motive force, through the arms and to the object being pulled (figure 14-13). If the elbows are bent during pulling, a shock absorbing spring has been added to the system that dampens the peak amount of force the system can produce. In the latter case, a completely extended elbow represents a condition where the bones act as a mechanical scaffolding thus reducing the amount of muscular force required to statically hold the object overhead (figure 14-14). Bending the elbows increases the exertional forces that must be produced by the elbow extensors to keep the load stable over head, a much more difficult condition to maintain for any period of time. So even though Grandma has always said to "put some elbow grease into it" when she wanted more effort and Drill sergeant bellowed "all I want to see is elbows and *@!+^%les" when he wanted more effort and less standing around, in pulling and supporting weight using the elbows is not the right thing to do.

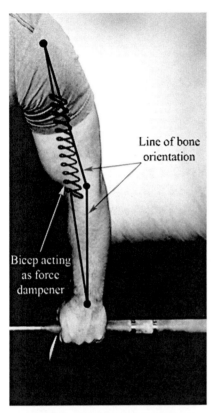

Line of bone
orientation

Bicep acting
as force
dampener

Figure 14-13. Incomplete extension of the elbow during any pulling movement reduces efficiency of the force transfer.

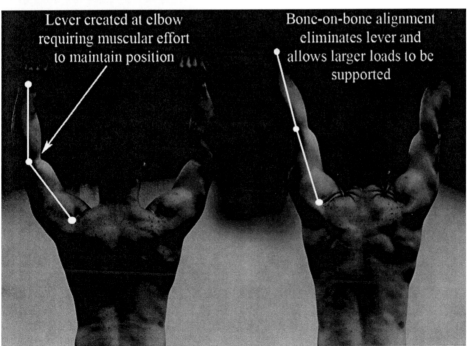

Lever created at elbow
requiring muscular effort
to maintain position

Bone-on-bone alignment
eliminates lever and
allows larger loads to be
supported

Figure 14-14. Incomplete elbow extension when supporting and overhead load reduces efficiency and performance.

"So you see, by applying the basic principles of the scientific method to the matter, we learn very quickly that the myth of the chupacabra is just that - utter crap. Now, if you apply the same principles to Catholicism, an interesting thing occurs ..."

*- **Dr. Thaddeus Venture***

15 - THE WRIST & HAND

The wrist and the human hand - complicated, capable, communicative, and uniquely human. Virtually every exercise and sporting activity, from putting on running shoes, holding onto a bar, and even pounding on another person (in the ring or octagon) is facilitated by the structure and abilities of the wrists and hands. The bony anatomy between the distal ends of the radius and ulna to the finger tips is comprised of twenty seven bones and lots of muscles. The radius is the twenty eighth bone of the system as the ulna does not articulate with the bones of the wrist (figure 15-1). Muscles that contribute to wrist and hand control can go as high up the system as having shoulder attachments or can be completely localized to the hand.

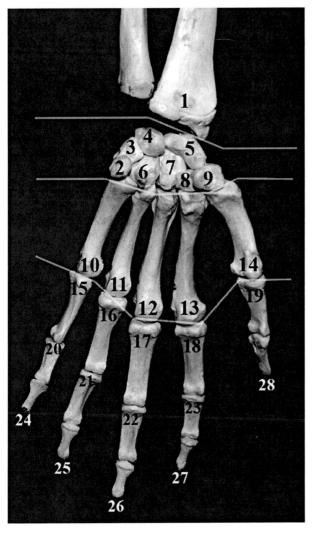

Figure 15-1. The bones of the wrist and hand (anterior or palmar side). 1 = radius, 2 through 9 = carpals, 10 through 14 = metacarpals, 15 through 28 = phalanx.

BONES

The architecture of the wrist and hand is constructed of many small little bones. This should make intuitive sense in that given the small and detailed movements required, small parts are necessary.

Radius - The distal end of the radius articulates with all but one of the bones in the proximal row of bones called the carpals (figure 15-1). A distal feature of the radius that is important is its widening and thickening, thus providing a larger surface for muscular attachments. At the distal end the radius is at least double the size as the ulna. Both the radius and ulna have defined pointy distal ends. The points are the *radial* and *ulnar styloid processes* (figure 15-2).

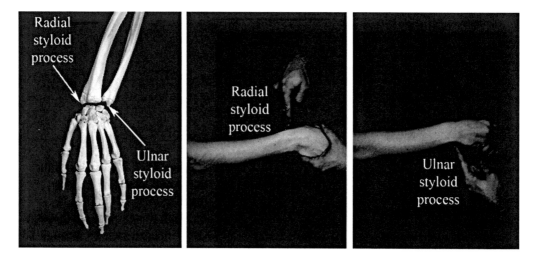

Figure 15-2. Finding the radial and ulnar styloid processes.

There are eight *carpal bones*, or carpi, arranged in two curved assymetrical rows, a (1) proximal row consisting of the scaphoid, lunate, triquetrum, and pisiform, and a (2) distal row consisting of the trapezium, trapezoid, capitate, and hamate (see figure 15-1). The proximal carpal row forms a sort of an arch that articulates with the radius, forming the wrist joint. The distal row forms the transition from wrist to hand. The carpals are irregularly shaped and are somewhat easy to identify (if you make sure you are looking at the correct - anterior or posterior - side of the skeleton). Some of the carpals are difficult to palpate because they are obscured by ligaments and tendons, but many possess characteristic protuberances and may be identifiable through palpation on the side with protruding features.

Scaphoid - The scaphoid bone is the largest and most lateral (thumb side) of the carpals in the proximal row (figure 15-3). It articulates with the radius proximally and with the trapezium and trapezoid bones in the distal row of carpals. The scaphoid can be palpated fairly easily. It is the bone on the lateral side of the wrist immediately distal to the end of the radius. It is flanked, to the anterior and posterior, by two tendons at the base of the thumb. Spread the fingers and lift the thumb to make the tendons prominent, then relax them in order to palpate the lateral surface of the bone. As an aside here, when the two tendons at the base of the thumb were tense and prominent, there was a small depression in between them. This little fossa is historically known as the *anatomical snuff box*. It was named such after the introduction of tobacco, particularly tobacco snuff, to Europe during the Renasissance. If a proper man-made snuff box was not available, snuff was placed into the fossa and inhaled through the nose. Tobacco was thought to be healthy and therapeutic for disease throughout much of human history.

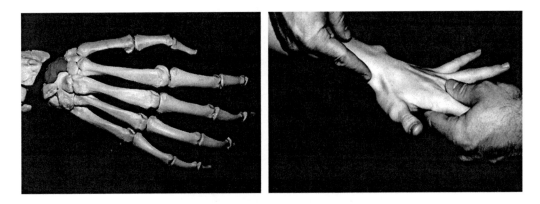

Figure 15-3. Posterior (back of hand) view of the scaphoid (left). Appearance of the anatomical snuff box during scaphoid palpation (right).

Lunate - The lunate bone, immediately medial to the scaphoid, has a very crude crescent shape hence the reference to the moon (luna) in its name. Physically it occupies the center of the proximal row of the carpals (figure 15-4. There are five articulations between the lunate and surrounding bones; to the distal radius, to the scaphoid and triquetrum bones of the proximal row of carpals, and to the capitate and hamate bones of the distal row of carpals.

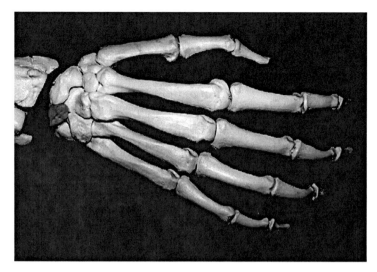

Figure 15-4. Posterior view of the lunate.

Triquetrum - The triquetrum bone in the proximal row of the carpals is roughly triangular shaped when viewed from the posterior. It lies between the lunate and pisiform bones (figure 15-5). Although it is very near the ulna it does not articulate with it, rather its articulations are strictly with other carpals, the lunate and pisiform of the proximal row and the hamate of the distal row. The triquetrum can be palpated by finding first the ulnar styloid process then radially deviating (abduct) the hand at the wrist. A bump will appear directly inferior to the unlar styloid process after deviation, the triquetrum's medial aspect.

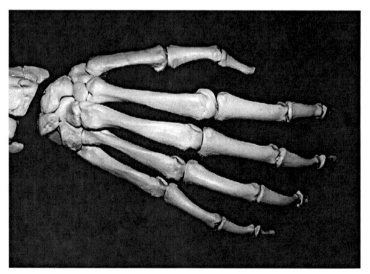

Figure 15-5. Posterior view of the triquetrum.

Pisiform - The pisiform is the fourth and most medial bone of the proximal row of carpals (figure 15-6). It is a very small bone that articulates only with the triquetrum. In fact, it is classified as a sesamoid bone, situated in the ligament of the truiquetrum. As the two are in such close proximity, they may be confused during palpation. The most proximal prominence is the triquetrum, the most distal and anterior (in front of the triquetrum) is the pisiform.

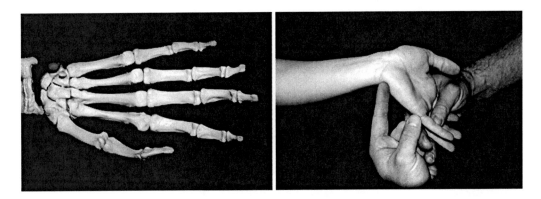

Figure 15-6. The pisiform can only be seen and palpated from the anterior (palm side).

Trapezium - The first bone in the second (distal) row of carpals is the trapezium. It sits between the scaphoid and the base of the first metacarpal (base of the thumb) (figure 15-7). It articulates with the first and second metacarpals, scaphoid, and trapezoid bones. The anterior surface of the trapezium presents the tubercle of the trapezium, which can be palpated on the palmar side of the wrist, just proximal to the base of the thumb (wiggle the thumb, the bump that does not move just below the base of the thumb is the trapezium).

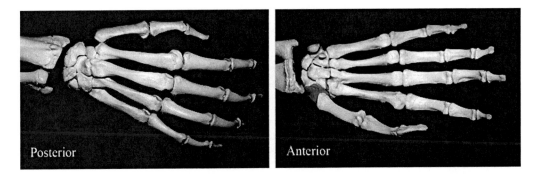

Figure 15-7. The trapezium from the posterior and anterior.

Trapezoid - On a skeleton the trapezoid is an easy bone to identify. It is the smallest in the distal row of carpals. While easy on a skeleton, its size and location make in unpalpable on a living human. It articulates with four bones; the scaphoid and the second metacarpal along the longitudinal axis and the flanking trapezium and capitate bones (figure 15-8).

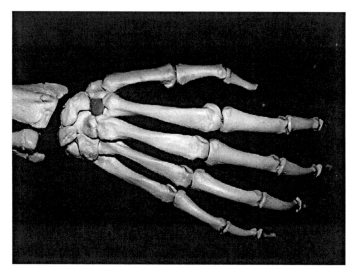

Figure 15-8. Posterior view of the trapezoid.

Capitate - The capitate is the largest of all the carpal bones and physically occupies the center space of the wrist (figure 15-9). With firm pressure applied just proximal to the posterior base of the third metacarpal (middle one), a portion of the capitate can be palpated. Aside from it being the largest, it has more articulations than all of the other carpals, having seven articulations; the scaphoid, lunate, trapezoid, hamate, and the second, third, and fourth metacarpals.

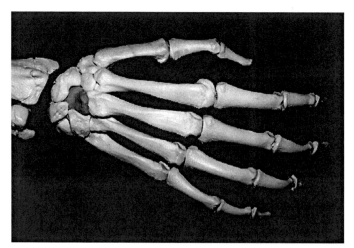

Figure 15-9. Posterior view of the capitate.

Hamate - The hamate is the most lateral bone in the distal row of carpals. It lies lateral to the capitate and crudely parallels the orientation of the combined triquetrum-pisiform pair just above it (figure 15-10). The hamate can be palpated on the anterior side of the wrist as it features a prominent protrusion called the *hamulus* (or more commonly the hamate hook) on its anterior aspect that lies just distal to the prominence of the pisiform. The thin structure of the hamulus predisposes it to occasional fracture from abrupt and forceful contact. In sport, it is seen occasionally in hackers (golfers who aren't that good) who have hit the ground with the club face not the ball. The torque created drives the club handle into the carpal, producing a hairline fracture. Fractures are also seen in baseball, generally as a result of catching a ball in the palm of the glove, not in the web. In baseball, unlike in golf, level of expertise is not a factor as there are numerous instances of this injury in professional players, some who had the hamulus removed in order to eliminate pain and potential for re-injury.

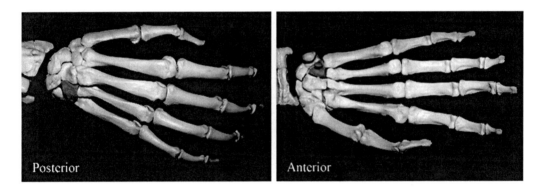

Figure 15-10. Posterior and anterior views of the hamate.

Metacarpals - The metacarpals are the long bones of the hand run between the distal carpals and the phalanges (figure 15-11). They form the palm and what we call the back of the hand. Their distal ends are the knuckles when we make a fist. There is a general characteristic shape to a metacarpal; concave proximal end, arched anterior surface, fairly straight posterior surface, and convex distal end (the head). The metacarpals are numbered, lateral to medial (thumb side in), from one to five. All five can be palpated from the posterior aspect.

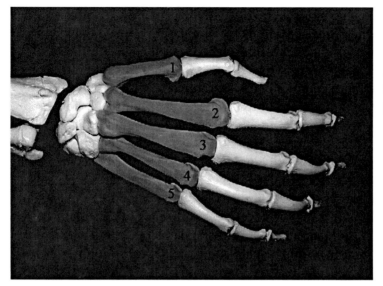

Figure 15-11. The metacarpals from the posterior.

Phalanges - The phalanges are also known as digits or the fingers. As with the metacarpals, they are numbered from one to five, lateral to medial (thumb to pinky). As noted parenthetically, there are other common naming conventions for the fingers that may help eliminate confusion:

    1 = thumb or pollux
    2 = index finger or fore finger or pointing finger
    3 = middle finger or bird
    4 = ring finger
    5 = little finger or pinky

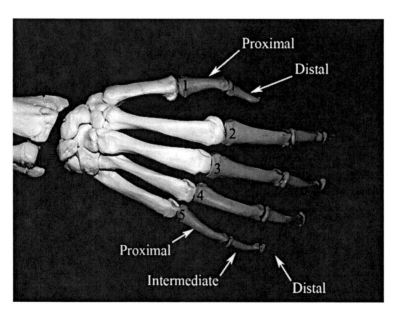

Figure 15-12. The phalanges are composed of a proximal, intermediate, and distal phalanx (posterior aspect).

Each phalange, a plural, is comprised of three phalanx (singular) (figure 15-12). The most proximal phalanx articulates with the distal end of the same numbered metacarpal and the proximal end of the intermediate phalanx. The intermediate phalanx articulates with the distal end of the proximal phalanx and proximal end of the distal phalanx. The distal phalanx only articulates with the intermediate phalanx as it represents the finger tip. The exception to this arrangement is phalange one, the thumb (or pollux - this alternate name becomes useful information later) which has no intermediate phalanx.

JOINTS

With twenty eight different bones present, there are numerous joints occurring throughout the wrist and hand. The wrist and carpal bones are attached to each other by a complex web of ligaments running between the radius and carpals, the ulna and carpals, and between the individual carpal bones. The joints generally can be palpated by locating the associate bones but the frequently overlying ligaments often obscure the joints, making palpation a practiced art. Some of the less important joints (all joints are important) have been omitted from presentation here.

Wrist - The wrist is the articulation of the radius, scaphoid, lunate, and triquetrum bones into a joint complex (figure 15-13). It can also be called the radiocarpal joint. The wrist is capable of moving in two anatomical axes, the sagittal and frontal - a biaxial joint. If you consider the kinetic chain of the wrist, the wrist is an intermediary that transfers force from the hand to the radius when lifting (pulling) an object OR transfers force from the radius to the hand when pushing objects. It is important to note that those forces are also transferred, by physical necessity, to the ulna through engagement of ligamentous tissues.

The major ligaments of the wrist are the *radial collateral* and *ulnar collateral ligaments* (figure 15-14). The radial collateral ligaments lies laterally in the ligament complex and attaches to the radial styloid process and to the scaphoid bone of the proximal row of carpals. It is a "collateral" ligament so a second strand that attaches to the trapezium and base of the first metacarpal is also present. It engages to limit ulnar deviation (adduction of the hand at the wrist). The ulnar collateral ligament is medial (on the ulnar side) and the two strands attach the ulnar styloid process to the triquetrum and pisiform carpals. This particular ligament acts to limit radial deviation. Both the radial and ulnar collaterals can be palpated with firm pressure during repeated deviation/relaxation cycles. Radial deviation will reveal the ulna collateral ligament and ulnar deviation will reveal the radial collateral ligament.

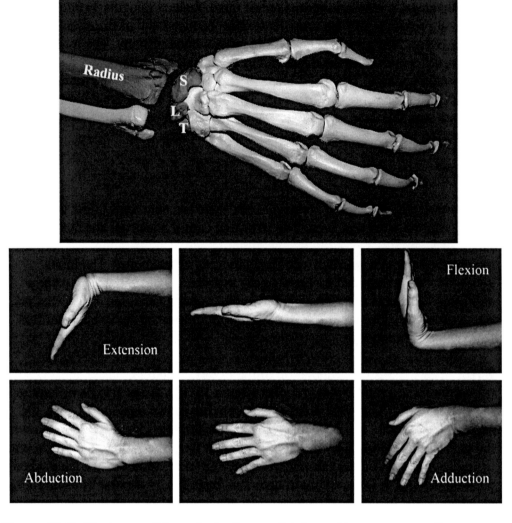

Figure 15-13. The bones of the wrist (radiocarpal) joint and its movements. S = scaphoid, L = Lunate, T = triquetrum. Abduction is also known as radial deviation, adduction as ulnar deviation.

On the posterior aspect of the wrist there is a group of deep and unpalpable ligaments call the *dorsal radiocarpal ligament* (figure 15-14). There are four defined segments of the ligament, all arising from the posterior lip of the radius just proximal to the joint. They will attach distally to the scaphoid, triquetrum, lunate, and capitate bones.

On the anterior aspect of the wrist there is a group of deep and unpalpable ligaments called the *palmar radiocarpal ligament*. There are six defined segments of the ligament, all arising from the anterior lip of the radius just proximal to the joint. They will attach distally to the scaphoid, triquetrum, lunate, and capitate bones (figure 15-15).

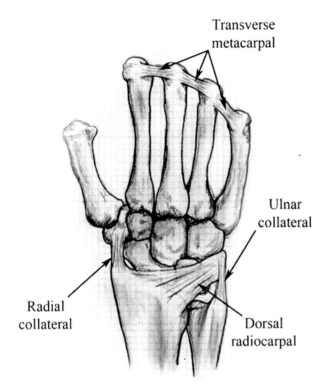

Figure 15-14. Posterior ligaments of the wrist.

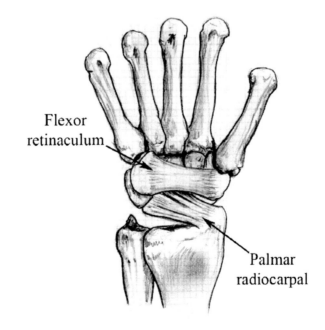

Figure 15-15. The flexor retinaculum and palmar radiocarpal ligament of the anterior wrist and hand.

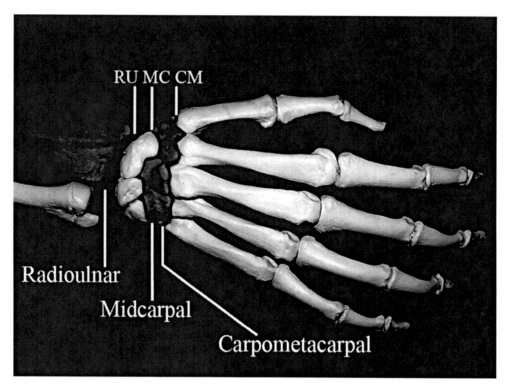

Figure 15-16. Orientation of the radioulnar, midcarpal, and carpometacarpal joints.

Intercarpal Joints - The intercarpal joints are the articulations between the individual carpal bones. Frequently you will see the intercarpal joints referred to as one of two collections of joints within the carpus, referring to the lateral and medial attachments between adjacent carpals. Midcarpal joints would then be attachments between the proximal and distal rows of carpals. These are planar synovial joints and although capable of limited movement between the carpals, they do contribute to wrist mobility. The carpals form an arch in the tranverse plane that is concave on the palmar side through which ligaments, nerves, and blood vessels pass. The arch deepens with flexion of the wrist and flattens with extension. This is why it is important to mildly extend the wrist when palpating for a radial pulse (checking the heart rate at the wrist).

The *flexor retinaculum*, also known as the transverse carpal ligament crosses over the carpal bones, forming a tunnel through which tendons and nerves can pass in a protected environment. As in the similar retinaculum in the ankle, it also prevents bowstringing of the tendons during contraction. It attaches to the pisiform and the hamate (along the hamulus) on the medial side and to the scaphoid the trapezium laterally (figure 15-15). It is contiguous with the extensor retinaculum on the dorsum of the hand.

The *extensor retinaculum*, also known as the dorsal carpal ligament, lies at a downward angle, lateral to medial, across the carpal bones and forms a connective tissue bridge over the extensor tendons, helping to keep the tendons in place. It has an attachment to the lateral border of the distal radius and to the triquetrum and the pisiform bones of the medial side of the wrist. You can look at the retinaculi as a guide for tendons, one that keeps them in a location that facilitates appropriate and mechanically efficient transmission of force. Combined, the flexor and extensor retinaculum form a robust band around the wrist at roughly the level of where a watch is worn.

Carpometacarpal Joints - These joints are easy to identify and describe, five of them occur where the distal row of carpals articulate with the proximal end of the metacarpals (figure 15-16). Each such articulation is a carpometacarpal joint. A small amount of pressure on a relaxed and flexed wrist joint allows palpation on the dorsal surface of hand. There is some debate regarding the number of ligaments present and their names. The five that seem to be most accepted are the anterior oblique, ulnar collateral, first intermetacarpal, posterior oblique, and dorsoradial ligaments.

Metacarpophalangeal Joints - The metacarpophalaneal joints are quite easily palpated. These are the condyloid joints or the "knuckles" where the metacarpals meet the phalanges. The rounded end of the metacarpal sits in the shallow bowl of the proximal phalanx (figure 15-17). These are biaxial joints allowing flexion, extension, abduction, and adduction of the phalanges. The exception here is the thumb which is a uniaxial joint, allowing flexion and extension. The joints are among the easiest to visualize and palpate. Simple make a fist to demonstrate the knuckles. Each joint can be palpated when the hand is relaxed. There are a number of ligaments present. One of the most interesting are the *transverse metacarpal ligaments* occurring between adjacent metacarpals (figure 15-14). These ligaments bind the heads of the metacarpals in close proximity, thus preventing excessive spreading and defining the maximum width of the distal palm. They can be palpated by gently squeezing the tissues just behind the webs between the fingers.

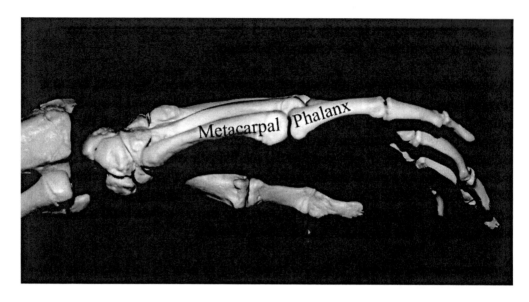

Figure 15-17. The metacarpophalangeal joint of digit 5 (medial view).

Interphalangeal Joints - The articulation of two sequential phalanx is an interphalngeal joint (figure 15-18). There are two such joints in each phalange with the exception of the thumb which has but one interphalangeal joint between its proximal and distal phalanx. The interphalangeal joint closest to the metacarpals is termed the proximal interphalangeal joint. The one farthest away is the distal interphalangeal joint. Each joint can be easily identified by simple curling the fingers up making palpation elementary. Each interphalangeal joint has several ligaments surrounding it to include cruciate (cross the joints) and collateral (down the sides) ligaments.

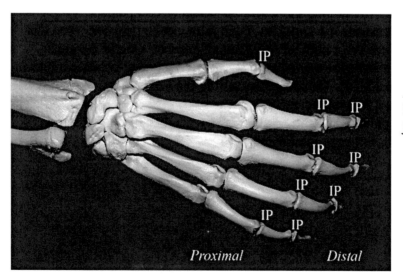

Figure 15-18. The interphalangeal joints.

There are a many many more small ligaments encasing the bones of the wrist, forming a network of interwoven connective tissue strands. This extensive system reinforces and holds the joints in close proximity while still allowing appropriate mobility.

MUSCLES

The muscular structure of the wrist and hand is quite complex, a function of the complexity of tasks it can and must perform. The extreme degree of manual dexterity requires many muscles that can be precisely controlled. The human wrist and hand fulfills this requirement with twenty sets of muscles acting on the joints to produce movement. Another functional necessity is that the muscles present must be able to generate significant force in order to capitalize on the human opposable thumb (for a great explanation of the importance of the human thumb, read "The Panda's Thumb" by the late Stephen J. Gould).

In general, there are a two levels of stratification used to lay out the muscles of the wrist and hand. The first strata is the division of the muscles into those that arise outside of the wrist and hand, the *extrinsic muscles*, and those that arise and end within the wrist and hand, the *intrinsic muscles*. The second strata is the division by function, flexor or extensor. Sometimes you will see further description as anterior/posterior or superficial/deep.

Extrinsic Flexors

Flexor carpi radialis - The name of this muscle can provide all you need to know. It is a flexor and since flexors are anterior, it must be on the anterior of the arm near the radius. It attaches distally to the second and third carpi (carpi is plural). It is attached proximally to the medial epicondyle of the humerus, on the radial side of the arm (figure 15-19). The basic actions driven by contraction of the muscle are flexion of the hand and, as the muscle angles across the forearm medial to lateral, it assists in abduction of the hand. Palpation of the tendon is fairly simple, flex the wrist and look for the closest tendon to the thumb on the radial side. By following the tendon back up toward the proximal attachment on the medial ulna, about half way up the forearm the cable-like tendon becomes softer, this is the beginning of the muscle.

Palmaris longus - This is a small muscle that may actually be absent in ten percent or more of the population. It has been reported in the clinical literature that its absence does not affect wrist flexion and grip strength. An interesting finding, as flexion of the wrist is its primary function. The muscle attaches proximally to the medial epicondyle of the humerus and distally to the proximal center of the *palmar aponeurosis*, a fanned sheet of fascia lying across the palm into which many ligaments and tendons invest themselves (figure 15-19). The tendon running between muscle and the aponeurosis is quite long. When the muscle contracts, the palmar fascia is tensed. The tendon can be palpated by placing the tips of the thumb and pinky together while flexing the wrist. The tendon, if the muscle is present, will present as a prominent tendon running longitudinally through the center of the superficial wrist. The palmaris longus lies immediately lateral to the flexor carpi radialis.

Flexor carpi ulnaris - Along the medial side of the palmaris longus, forming the contour of the medial forearm, you will find the flexor carpi ulnaris. As one would expect from its name, flexor indicates both its function and location on the anterior side of the forearm, carpi indicating an attachment to at least one carpal, and ulnaris suggests a muscle associated with the ulna. The flexor carpi ulnaris does attach proximally to the medial side of the olecranon but it also has a proximal attachment to the lower aspect of the medial epicondyle. It attaches distally to the anterior surface of the pisiform bone (figure 15-19). The muscle's anterior location makes it a flexor, its medial attachments also make it an adductor of the hand. Palpation of the tendon is possible at a point just proximal to the pisiform during wrist flexion. The fingers can then be walked up the medial side of the forearm during flexion and relaxation to isolate the muscle mass.

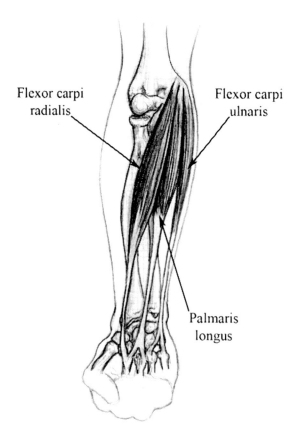

Flexor carpi radialis

Flexor carpi ulnaris

Palmaris longus

Figure 15-19. The medial, anterior, and superficial flexors of the wrist and hand.

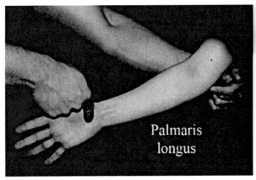 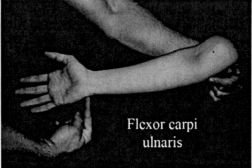

Palmaris longus

Flexor carpi ulnaris

Figure 15-20. Finding the superficial flexors is a matter of finding the proximal attachment and either the distal attachment or the region through which their tendons pass. Then making a fist or flexing the wrist should reveal them.

Flexor digitorum superficialis -This is a finger flexor as the name would attest, but it is not so superficial as to be easily palpated. It lies beneath the brachioradialis, pronator teres, flexor carpi radialis, and the palmaris longus. The flexor digitorum longus proximally attaches at the distal radius and medial epicondyle of the ulna (figure 15-21). Distally, it divides into four tendinous splits and attaches to the intermediate phalanx of phalanges two through five. It is a flexor, active in wrist flexion and in flexion of the fingers, as in making a fist.

Pronator quadratus - This is an anterior muscle that is not a flexor. It is, as the name implies, a pronator of the hand. It attaches along the distal aspect of the last few inches of both the radius and ulna (figure 15-21). The direction of the fiber orientation is lateral across the space between the two bones.

Flexor pollicus longus - Here is where the Latin nomenclature for the thumb comes into play. Pollux means thumb, so the flexor pollicis longus should be a flexor of the thumb. Attaching proximally to the upper half of the radius along with an attachment to the coronoid process of the ulna. It attaches distally to the base of distal phalanx of the thumb (figure 15-22). The muscle can be identified by palpating the lateral aspect of the upper forearm during repeated flexion/relaxation cycles of the thumb against the fore finger (pinching). The movement of the tendon and muscle should be visible and palpable.

Flexor digitorum profundus - The name flexor digitorum profundus suggests a flexor of the fingers. The other word in the name, profundus, sounds a lot like profound. A profound thought is a deep thought. In Latin, profundus means deep or profound, so the muscle should be a deep one that is not likely palpable. The

251

muscle has a long proximal attachment along the ulna, spanning from near the olecranon down about two thirds its length. The distal attachments of the profundus are interesting and relatively unique. Instead of a single tendon emerging from the muscle and then dividing into slips for multiple attachment, there are four tendons that arise independently along the distal border of the muscle belly. The tendons then continue on to attach distally to the bases of the distal phalanx of digits two through five (figure 15-22).

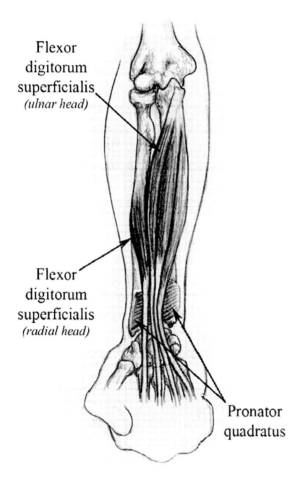

Flexor
digitorum
superficialis
(ulnar head)

Flexor
digitorum
superficialis
(radial head)

Pronator
quadratus

Figure 15-21. The mid-layer flexor digitorum superficialis and the pronator quadratus.

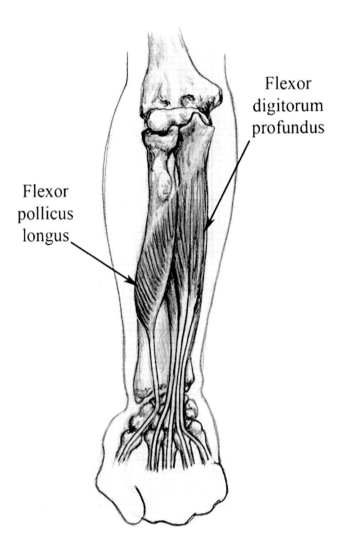

Flexor digitorum profundus

Flexor pollicus longus

Figure 15-22. The deep anterior flexors, the flexor pollicus longus and flexor digitorum profundus.

Extrinsic Extensors

Extensor carpi ulnaris - This extensor also attaches proximally to the humerus at the lateral epicondyle. It then runs down the medial border of the forearm where it has a considerable attachment to the ulna, crosses the wrist, and attaches distally to the base of the fifth metacarpal (figure 15-23). This orientation also makes it an adductor of the wrist and hand. Palpation is fairly easy by finding the two attachment sites and observe/palpate the expected line of the muscle along the forearm while performing simultaneous wrist flexion and ulnar deviation. It should be visually apparent and palpable with repeated contraction-relaxation cycles.

Extensor digitorum - This muscle has essentially opposite actions and attachments as the flexor digitorum profundus. It too attaches proximally to the lateral epicondyle of the humerus and distally to the first phalanx of the second through fifth phalanges, but on the posterior of the hand (figure 15-23). The tendons of this muscle are very apparent on the back of the hand and wrist during extension of the fingers.

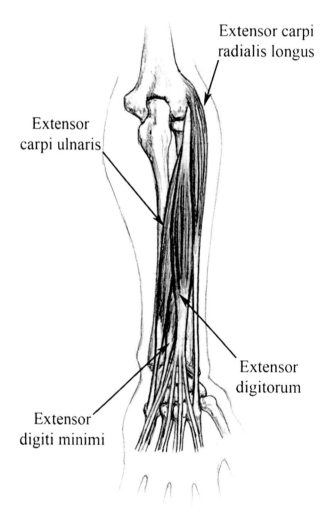

Extensor carpi radialis longus

Extensor carpi ulnaris

Extensor digitorum

Extensor digiti minimi

Figure 15-23. The most superficial of the wrist and hand extensors (posterior view).

Extensor digiti minimi - This is a very small extensor muscle appearing slightly medial to the extensor digitorum. It attaches proximally to the lateral epicondyle of the humerus and distally to the distal phalanx of the fifth phalange (figure 15-23). Extension of the fifth phalange will make the distal tendon prominent for palpation.

Extensor carpi radialis longus - The extensor muscles of the forearm are on the posterior or dorsal side. As seen before, carpi indicates an attachment to one or more carpal bones, radialis indicates proximity to the radius, and longus suggests a long muscle. The extensor carpi radialis longus attaches proximally to the supracondylar ridge above the lateral epicondyle of the humerus, an attachment obscured from palpation by the overlying triceps brachii. The muscle then becomes palpable as it runs parallel and inferior to the brachioradialis then attaches distally to the base of the second metacarpal (figure 15-23). As the muscle orientation runs along the lateral aspect, it can also act as an abductor of the wrist and hand (radial deviation).

Extensor carpi radialis brevis - The similarity of names, indicates a muscle similar to the extensor carpi radialis longus, just a shorter muscle. The major differences between the two is that much of the brevis attaches proximally below the longus - to the lower aspect of the lateral epicondyle of the humerus - and the brevis attaches distally to the base of the third metacarpal (figure 15-24). It shares the same functions at the longus.

Abductor pollicus longus - The abductor pollicis longus is a deep muscle that runs from its proximal attachment along the mid-dorsal aspects of the radius and ulna to a distal attachment at the base of the first metacarpal (figure 15-24). It acts as an abductor of the thumb.

Extensor pollicus longus - The extensor pollicus longus is a deep muscle, mostly obscured from palpation by the abductor pollicus longus. It attaches proximally along the radius near the abductor pollicus longus and attaches distally to the base of the distal phalanx of the thumb (15-24). It acts to extend the distal phalanx. Its orientation also enables it to assist in wrist abduction. The tendon of the muscle can be palpated at the anatomical snuff box, it is the medial tendon bordering the depression.

Extensor pollicis brevis - The brevis lies below and essentially parallel to extensor pollicus longus. Its proximal attachment is just below the longus and attaches distally just short of the longus, at the base of the first phalanx of the thumb (figure 15-24). Its function is to extend the first phalanx of the thumb. The tendon of the muscle can be palpated at the anatomical snuff box, it is the lateral tendon bordering the depression.

Extensor indicis - The extensor indicis as suggested by the name is a posterior forearm muscle, an extensor. Indicis refers to the index finger. The muscle proximally attaches to a small area approximately three fourths the way down the ulna and attaches distally to the distal phalanx of the second phalange (figure 15-24). It extends the index finger. The distal tendon is easy to palpate during repeated extension and relaxation of the index finger. The tendon and the movement can be followed to the muscle, it should appear as the tendon crosses the wrist above phalange three.

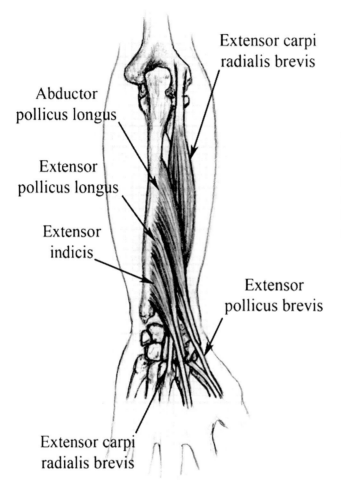

Abductor pollicus longus

Extensor pollicus longus

Extensor indicis

Extensor carpi radialis brevis

Extensor pollicus brevis

Extensor carpi radialis brevis

Figure 15-23. The deep extensors of the wrist and hand (posterior view). Note that the tendon of the extensor carpi radialis brevis runs under the other muscles depicted.

The *intrinsic muscles* are the abductor pollicis brevis, opponens pollicis, flexor pollicis brevis, adductor pollicis (thenar muscle) and opponens digiti minimi, flexor digiti minimi brevis, abductor digiti minimi (hypothenar muscle). The thenar muscle is the meat of the hand below the thumb, the hypothenar is the meat of the hand below the pinky. Although these muscle provide fine motor control primarily, they can be developed for improved strength or endurance through appropriate exercise.

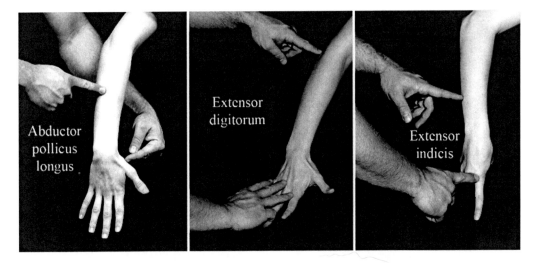

Abductor pollicus longus

Extensor digitorum

Extensor indicis

Figure 15-25. Locating the superficial flexors is a matter of finding the proximal attachment and either the distal attachment or the region through which their tendons pass. Then making extending the fingers and wrist should reveal them. Even some of the underlying muscles can be localized by doing the same an feeling for movement or tension in their tendons.

MOVEMENTS

The joints and muscles of the hand produce the same movements as other joints, flexion, extension, abduction, adduction, rotation, and circumduction. But the sheer volume of joints working provides for exquisite control of the orientation of the hand and fingers in space.

At the wrist the movements possible are flexion, extension, adduction (ulnar deviation), abduction (radial deviation) and circumduction (figure 15-26).

At the fingers there can be flexion at one or more joints, extension at one or more joints, adduction at the metacarpophalangeal joint, abduction at the metacarpophalangeal joint, and circumduction at the metacarpophalangeal joint (figure 15-27).

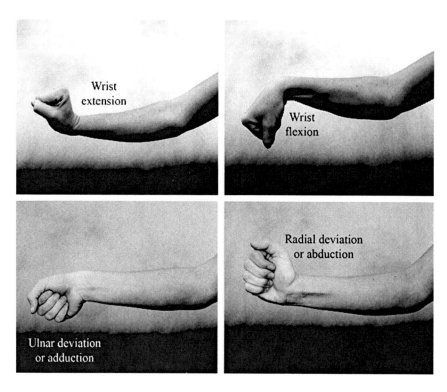

Figure 15-26. The movements of the wrist joint.

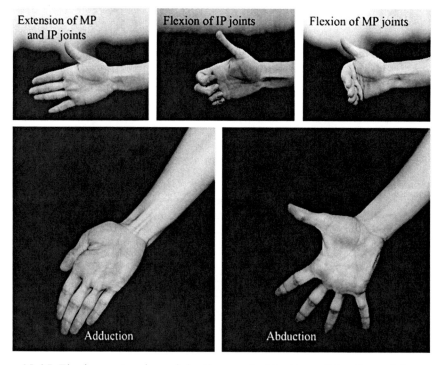

Figure 15-27. Flexion, extension, abduction, and adduction of the finger joints.

At the thumb, a unique movement besides flexion, extension, adduction, abduction, and circumduction is possible, opposition (figure 15-28). Opposition of the thumb is a unique ability of humans, discriminating us from other vertebrates. In opposition, the thumb is moved toward the palm to "oppose" the remaining fingers. This movement allows us to pinch, grasp, and hold (wrap the hand around an object). Reposition is the opposite of opposition and is simple the return of the thumb to anatomical position. In the course of human development it has been the human ability to hold and manipulate tools that moved the human race forward.

It is almost universally thought that the development of the human thumb was driven by the need to create tools for human use. The newest anthropological data suggests that the human thumb arose about six million years ago, more than three million years before the earliest evidence of man-made tools. That is leading to an emerging hypothesis that intellectual development rather than physical ability drove the development of tools. This is supported by the use of a tool (a thin stick) to harvest termites from their mounds as a food source by some modern primates. An ability to use an existing tool is present in some primates, however they lack the intellectual capacity to create new tools for specific purposes.

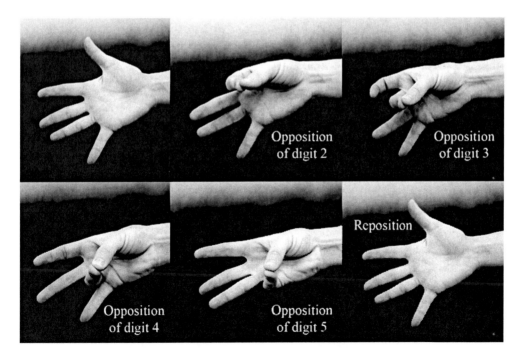

Figure 15-28. Opposition of the thumb against flexion of individual digits.

A Gripping Situation

Grip strength is an important aspect of many sports. Whether gripping an implement - bat, ball, bar, ring, to name just a few - or gripping another human, a strong stable grip is a frequent necessity. The largest muscle masses providing grip strength are the extrinsic flexors of the wrist and hand. The most common exercise prescription for developing these muscles is the performance of wrist curls and wrist rollers. These exercises are usually prescribed in higher repetitions, often performed to failure. This is endurance training, not strength training. This is a glaring problem as grip strength can be strongly developed without such isolation movements that require relatively light weights thus limiting strength gain potential.

A solution to this problem is fairly straight forward, having two parts. The first part is to include exercises requiring the holding of heavy weights in the hands, such as deadlifts and cleans. In these exercises, the extrinsic flexors isometrically contract and hold the weight in the hands, a very strong training stimulus. The type of grip used does affect the amount of isometric work done by the flexors. A standard (double overhand) grip requires the most work as it is only the force of flexor contraction holding the bar in place - it is a simple muscle driven grip (figure 15-29). A powerlifting deadlift (alternating hand direction) grip reduces the amount of work required because as the bar rotates up in one hand as it rotates down in the other - an alternative direction force driven grip (figure 15-29 inset). You can lift more weight with deadlift grip than a standard grip, it removes, to a degree, grip strength as a limiting factor. There is another grip used in lifting a bar from the floor, the hook grip. In the hook grip, the thumb is adducted across the bar to be lifted then the fingers are wrapped around the bar and over the thumbs (figure 15-30). This reduces the amount of flexor strength needed to hold a weight in the hands - a friction driven grip. For grip strength development, the order of desirability is (1) the double overhand grip, (2) the deadlift grip, (3) the hook grip. An interesting way to work grip is to use a bar that the hand cannot close around thus making the flexor musculature the only means of keeping the bar elevated (figure 15-31). This can be done with chin/pull-ups on a fat bar or with deadlifts with a fat bar (It is a curious observation that you cannot press as much weight with a fat bar as you can with a standard bar - is grip involved?).

The second, follow-on, part of effective grip development is to include dynamic finger flexion into the training. There are any number of weighted and spring-loaded devices available to accomplish this task. One important thing to remember is that progression in load (making it heavier or more difficult), whether on a bar or a grip trainer, is always a necessity.

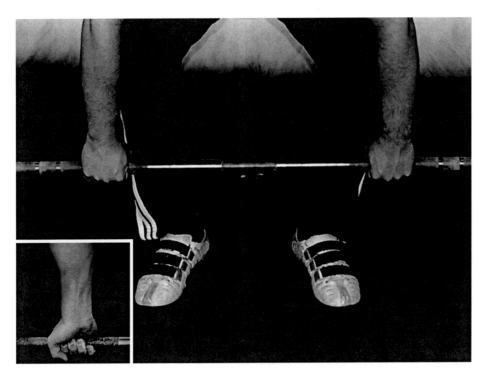

Figure 15-29. Standard double overhand grip where both hands are pronated. For a deadlift grip, one hand is supinated (inset).

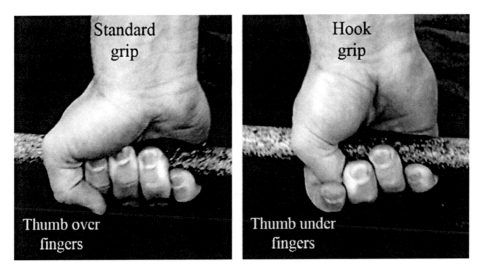

Figure 15-30. A standard or conventional grip where the fingers are wrapped around the bar and the thumb is laid over the fingers (left) offers less grip security than the hook grip where the fingers are wrapped over the thumb (right).

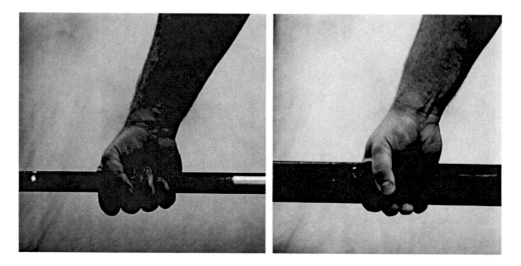

Figure 15-31. When using a 2 inch diameter fat bar (right), the force of friction is removed as an aiding factor and the angle of force application in the individual joint lever systems in the fingers make for a very difficult grip exercise.

As an added benefit, the intrinsic flexors are developed with the same strategies used with the extrinsic flexors, and at the same time. This has application to the real world weight room where there may be limited time to accomplish desired results. By developing the grip doing exercises, like the deadlift or clean, ones that also develop major muscle groups, there are better holistic gains in strength in less time. It also makes programming simpler, a perk for any coach or fitness trainer. The addition of the specialized flexion exercises would not occur until there was a specific trainee desire or if there was a functional need such as in the training of an alpinist (climber) or a strongman athlete.

16 – THE HEART & LUNGS

Everyone has a heart. Regardless if they are considered to be a heartless troll or a soft-hearted goody two shoes, everyone possesses in their chest a physically functioning heart that is architecturally the same ... with some normal and minor human variation of course. The majority of the heart is comprised of muscle and enclosed spaces wrapped in the muscle present. The arrangement of the muscles around enclosed spaces is important as the heart is a biological fluid pump, pumping blood to all points of the body having blood vessels. It pumps blood 24/7 from birth to death. If we consider the present lifespan to be 76 years, that works out to 2,796,192,000 pumping actions (beats) over the lifespan. That is a lot of work. The energy required to drive blood through the vessels has been calculated to be 0.0023 calories per each heart beat. This equates to roughly 6,518,900 calories used over the lifespan just to pump blood. Not that this would be typically considered "heart healthy" but it takes just about the energy content of one McDonalds [TM] hamburger each day to power the heart.

The human heart is not a terribly large organ, tipping the scales at between half a pound and three fourths of a pound (250 to 350 grams). In general the size of the heart is related to the size of the body containing it. A small body has a heart on the smaller end of the scale of norms and a large body has a heart on the larger end of the scale. For example a rat heart is about the size of the end of your pinky, an elephant heart is bigger than a basketball. A very quick and easy, and somewhat accurate way to visualize the size of your heart is to simply make a fist (figure 16-1). Mind you, this is not an absolute and accurate measure of dimension, it is a ballpark estimation.

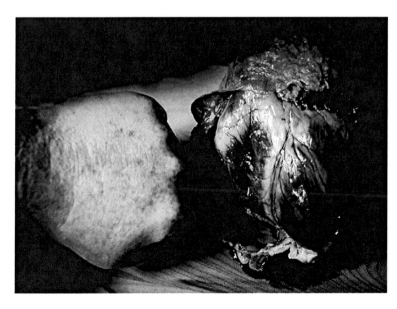

Figure 16-1. The heart is roughly the same size as the fist of the body it came from. Here a heart from a 200 pound human analog (pig) is about the same size as a fist of an 180 pound human. *Note:* a pig's heart is essentially identical in structure to a humans.

The human heart lies at about a 20-degree angle (downward, right to left) just posterior to the sternum, somewhat centered within the ribcage (a little biased to the left). The big blood vessels coming out of its top will be towards the right and the narrowed tip of the heart will be towards the left. Its lowest excursion is above the level of the xyphoid process. (figure 16-2). Further, it is nestled in between the lungs, liver and diaphragm, and endowed with some shock absorbing and metabolically available adipose deposits (fat). It is a well protected organ.

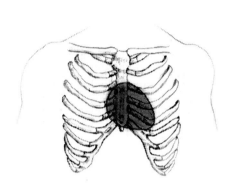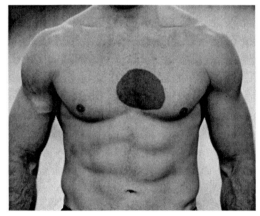

Figure 16-2. Orientation of the heart within the thoracic cavity (chest).

PROTECTION & LUBRICATION

The heart muscle is not just stuck out in the middle of other organs and such in the chest cavity. It is enclosed and bounded within a double-walled sac structure named the pericardium. The outer *fibrous pericardium* is of a very robust construction and provides for three basic functions:

- Protection of the heart from abrasion, puncture, and infection

- Service as an attachment point to surrounding anatomical structures

- Due to the limited extensibility of the pericardium, it limits the degree to which the heart can expand as it fills with blood.

The inner *serous pericardium* exudes pericardial fluid between itself and the outside of the heart muscle that acts as a lubricant for the heart during its contractile activity.

MUSCULAR ORGANIZATION

The heart is divided first into superior (upper) and inferior (lower) segments (Note: I use the term "segments" here as the heart is not symmetrical and the term "halves" is not appropriate). The upper segment contains two small sized chambers called *atria* (singular = atrium), each with an entry opening and an exit opening. The atria are fairly thin and look reminiscent of very small and partially deflated whoopee cushions, but they do have muscular walls (figure 16-3 and 4). The function of atria is to accept blood from the periphery then forcefully move it into the lower segment of the heart. The short distance and large diameter openings through which it moves blood allows for the light musculature present to function quite well. Both atria are subject to essentially the same work stress and have very similar structure.

Figure 16-3. The exterior of the heart from four perspectives; anterior, left lateral, posterior, right lateral (from left to right). Note the size difference between the atria and ventricles.

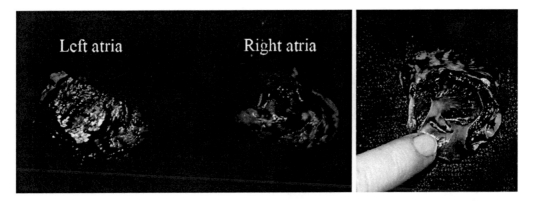

Figure 16-4. Isolated atria (superior view). Note, at the far right, the small size of the chamber and the thinness of the atrial wall.

The lower segment contains two much larger chambers called *ventricles*. Ventricles also have entry and exit openings but are much more heavily muscled than the atria. The thicker musculature is necessary, as both ventricles must produce higher forces than the atria due to higher pressures being required to move blood great distances and through smaller peripheral openings and vessels.

The heart is also divided into left and right segments, containing one atria and one ventricle (figure 16-5). This is a functional and morphological division. The right segment – the right atria and ventricle – are considered the pulmonary side, receiving oxygen depleted blood from the body and routing it to the lungs for oxygenation. The left segment – the left atria and ventricle – are considered the systemic side, receiving oxygenated blood from the lungs and delivering it to the body (all systems, hence means to remember "systemic circulation"). The right and left ventricles share a common muscular wall termed the ventricular septum (figure 16-6). The remaining muscular walls are called ventricular free wall. The free wall of the left ventricle is much thicker than that of the right ventricle. This is a functional consequence of the left ventricle muscle having to produce more force to move blood through the systemic circulation. The average arterial pressure needed to move blood to all parts systemic is about 100 mmHg (same units as barometric pressure), whereas only about 25 mmHg is required for pulmonary circulation.

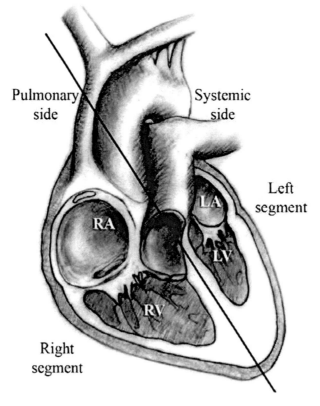

Figure 16-5. Functional division of the heart into right and left segments.

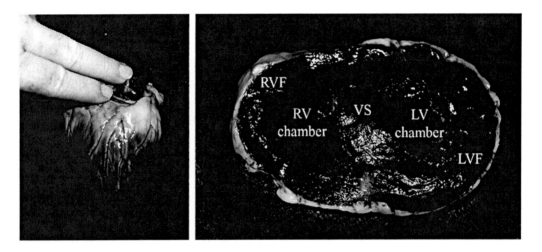

Figure 16-6. The exterior of a ventricle (left). A transverse cut across the heart produces a coronal view demonstrating the differences between the left and right ventricles. RV = right ventricle, RVF = right ventricular freewall, LV = left ventricle, LVF = left ventricular freewall, VS = ventricular septum.

The wall of the heart is composed of three distinct layers (figure 16-7). The outermost layer is in close communication with the inner layer of the pericardium and is called the *epicardium*. The innermost layer is the *endocardium*, a fairly thin but important layer as it contains the innermost lining of the heart chambers, the *endothelium* (a cell type). The endothelium produces and secretes a large number of bioactive chemicals important for circulatory function. The middle layer between the epicardium and endocardium is the *myocardium*, named such as it contains muscle, the bulk of the wall.

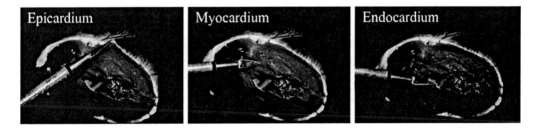

Figure 16-7. The layers of the heart wall.

The muscle present in the heart differs anatomically and functionally from skeletal muscle as studied earlier. Cardiac myocytes differ from skeletal myocytes in:

Morphology – more jigsaw puzzle piece looking, not fusiform

Intracellular communication – many intercalated discs present

Nucleation – mononucleated rather than multinucleated

The unique morphology of cardiomyocytes increases the communicative surface area and increases the number of adjacent cells with which they are in contact (figure 16-8). Combined with the presence of *intercalated discs*, little pores between two adjacent cells that allow free passage of chemical signals, simultaneous muscular contraction of a segment of the heart is possible. When a cell or set of atrial muscle cells is stimulated to contract, the communicative signal cascade made possible by these features drives all atrial muscle cells to contract. The same situation occurs in the ventricles.

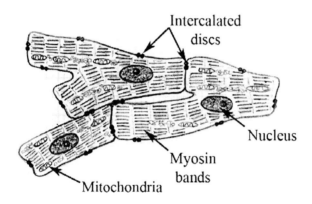

Figure 16-8. Schematic of cardiac muscle cells.

The actions of the atrial segment and the ventricular segment are sequential, the atria contracts first, then the ventricles. To make this ordered arrangement occur requires a set of specialized electro-conductive cells to be present.

CONDUCTIVE PATHWAY

Within the walls of the heart there exists a specialized set of cells that conduct and direct electrical impulses through the heart, from atria to ventricle, in an organized manner. The organized electrical activity induces a sequential muscle contraction that squeezes blood through the heart without allowing backflow. The electrical activity driving muscle contraction can be recorded and visualized, it's the electrocardiogram (ECG or EKG) that you have as part of your physical checkup or see on your favorite medical TV show.

There are seven commonly accepted components to the cardiac electro-conductive pathway (figure 16-9); sinoatrial node, internodal tract, Bachmann's bundle, atrioventricular node, bundle of His, right & left bundle branches, Purkinje fibers.

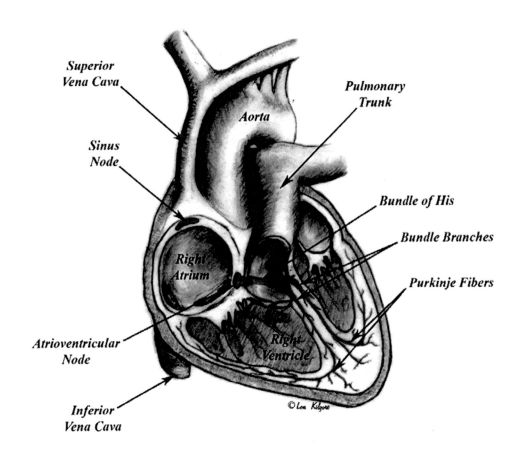

Figure 16-9. Basic elements of the cardiac conductive pathway. The internodal tract between the sinoatrial and atrioventricular nodes as well as Bachmann's bundle are not shown.

All of these specialized groups of cells arose originally from cardiomyocytes during development. It should then be no surprise that cardiomyocytes possess a unique electrical capacity. All cardiac muscle is autorhythmic, meaning it can generate its own electrical stimulus to drive contraction. If the normal electro-conductive pathway fails, a cardiomyocyte, or group of them, can take over and keep the heart beating – albeit with a greatly reduced efficiency. And yes, this means that in all those sci-fi and horror movies you've seen, the heart still beating after it has been precipitously and involuntarily removed from the thoracic cavity of an unfortunate character does have a basis in reality.

In a normal and healthy heart, the stimulus to begin the cardiac cycle of contraction begins at the *sinoatrial node*. Located in the superior right atrial wall, the sinoatrial node is considered the physiological pacemaker of the heart, setting the tempo of the heart rate. When someone has an "artificial pacemaker" surgically implanted, it is generally because their anatomical sinoatrial node is failing to produce a proper heart rate or is producing irregular rhythms. The sinoatrial node stimulates both atria to contract. Its location within the right atrium driving its contraction and the node's connection to the left atrium via *Bachmann's bundle*, passes electrical activity to the left atrium to drive its simultaneous contraction. Electrical impulse also traverses, superior to inferior, the length of the right atrium by way of the *internodal tract*. The internodal tract conducts the impulses generated at the sinoatrial node to the atrioventricular node. Note that all of the anatomical features depicted in figure 16-9 are not identifiable by simple visual inspection, a microscopic examination is required.

The *atrioventricular node* lies fairly superficially in the inner surface of the inferior right atrial wall, as short distance from intraventricular septum. The atrioventricular has a similar structure containing the same type of specialized conductive cells as the sinoatrial node. The atrioventricular node has a very distinguishing functional feature not seen in its superior counterpart, a fibrous sheath covering its surface. This sheath is not as conductive as the node itself, and delays the conduction of the electrical impulse. It acts as a sort of anatomical timer, allowing atrial contraction to occur before the ventricles receive their starting impulse. Once the impulse clears the sheath and reaches the atrioventricular node, the diameter of the subsequent conductive tissues is larger than that in the atria and as a result conducts electrical impulses more rapidly.

The atrioventricular node leads next to the *Bundle of His* leading from the node to the intraventricular septum. The bundle turns inferiorly and penetrates the septum. There it divides into the *right and left bundle branches* that direct themselves and their offshoots to the right and left ventricles respectively. These offshoot branches are called *Purkinje fibers* and they branch and divide extensively down the length of the septum then up and around the ventricular freewalls enveloping the ventricular muscle in a conductive network. This creates and extensive and nearly instantaneous spread of contractile stimulus enabling the large ventricular muscle mass, both right and left, to contract as a single unit. With the sequential and separate contraction of the atria and ventricles, the blood can then be moved in a controlled and directional manner that is responsive to biological demand.

BLOOD FLOW PATHWAY

As noted earlier, the right side of the heart receives deoxygenated blood from the body into the right atrium. The blood is delivered by the superior and inferior vena cava and collects into the coronary sinus, a sort of juncture between the superior and inferior vena cavas, and passes through the semicircular *Thebesian valve* into the atrium. After atrial pressure changes and the Thebesian valve closes to prevent regurgitation (backward blood flow, not barfing), the blood is pushed through the *tricuspid valve* into the right ventricle. The tricuspid can be identified by the three (tri = three) leaflets of which it is constructed. Blood ejection from the right ventricle occurs through the *pulmonary valve* after intra-ventricular pressure pushes it open. The blood from the right ventricle is destined for oxygenation in the lungs via the pulmonary artery (the *pulmonary circulation*) so the name of the valve should be easy to remember. Once the blood clears the lungs it returns to the heart by way of the pulmonary vein. It is worthwhile to note that the pulmonary artery and pulmonary vein are exceptions to the rule that arteries carry oxygenated blood away from the heart and veins carry deoxygenated blood to the heart. There have opposite roles to their naming convention. The oxygenated blood enters directly into the left atrium. Once full, the pressure within the left atria forces open the *bicuspid valve* and blood moves into the left ventricle. The bicuspid valve is identified by the presence of two leaflets. Rising pressure in the full ventricle shuts the entry valve and opens the *aortic valve* and the blood leaves the heart through, not surprisingly, the aorta.

The aorta is the most important artery in the body, in terms of blood delivery, after emerging from the heart it arches back towards the vertebral column and becomes know as the thoracic then abdominal aorta as it passes through those cavities. All the important appendicular arteries (like the subclavian and iliac arteries) and visceral arteries (like the hepatic and mesenteric arteries) are derived from the aorta. The majority of the blood leaving through the aorta is destined for the systemic circulation but a portion of it is immediately redirected from the aorta into the coronary circulation.

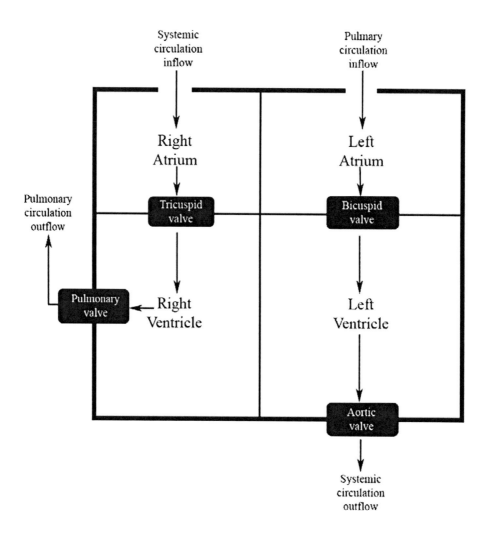

Figure 16-10. Schematic of blood flow through the heart valves.

The *coronary circulation* refers to an important and dedicated set of arteries and veins that perfuse the heart (figure 16-11). Coronary arteries diverge from the aorta as soon as the aorta emerges from the cardiac mass and then spread out along the surface of the heart and divide and penetrate down into the heart to provide an extensive blood flow to all parts of this essential muscle. There are two divisions of arteries, the left and the right. A *left common coronary artery* emerges from the aorta and divides into *the left anterior descending coronary artery* and the *left circumflex coronary artery*. The anterior descending artery runs, as you would expect, down the front of the left ventricle. The circumflex artery runs around the superior aspect of the left ventricle (anterior to posterior) just below the atrial juncture. Both of these arteries branch extensively to feed the left side of the heart. The right side of the coronary circulation is slightly less extensive. The *right common coronary artery* emerges from the aorta and runs anterior to posterior along the right superior surface of the right ventricle just below the right atria. A large artery, the *right posterior descending coronary artery*, divides off of the right common and both continue to divide extensively to perfuse the right side of the heart.

The coronary arteries are of significant interest to clinicians and of moderate interest to coaches and fitness trainers. Sedentary lifestyles are associated with the development of blockages of the coronary vasculature. The most common coronary artery disease and cause of blockage of these arteries is atherosclerosis. About a million people per year have reportedly died from causes related to coronary artery disease. This is a significant medical and social expense and a great deal of effort and funding has been directed at reducing the rate of mortality associated with coronary artery disease. Unfortunately the rate of death from coronary artery disease has been stable at about 900,000 to one million for about two decades. This indicates very little progress, a scientific work in progress so to speak. One thing that is certain is that every coach and every fitness trainer will work with someone, or many people, with undiagnosed atherosclerosis (it is the "silent killer"). A functional knowledge of first aid, established emergency plans, and continual surveillance of trainees during exercise is a necessity for all sport and exercise professionals. On the up side of coronary artery disease – continually active individuals are far less likely to develop atherosclerosis over the lifespan.

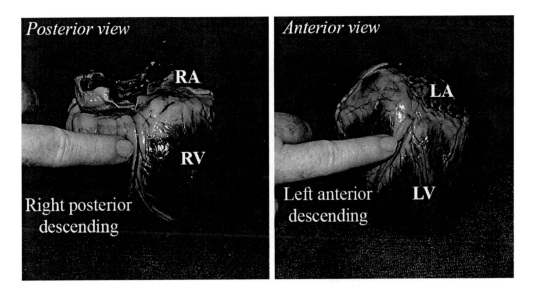

Figure 16-11. The most easily identified coronary arteries, the right posterior and left descending coronary arteries. Note the extensive branching of smaller arteries off of the main vessels.

WHAT IS BEING PUMPED?

When considering the anatomy of a pump, it is prudent to consider the anatomy of what it pumps. In this instance we need to examine the components of blood.

At the most broad of levels, blood is comprised of solids and liquid. The basic function of the blood is to transport dissolved or suspended materials to a location of consumption or a location of disposal.

The liquid portion of the blood is pretty straight forward in identity, water. Water with many mineral ions and bioactive molecules dissolved in it called plasma. Sort of a salt water solution with extras, in a very simplified view. The solid portion of blood is made of cells, platelets, cell fragments, and very large molecules. The ratio of these two components, solid and liquid, is expressed as a percentage and is called the *hematocrit*. We can easily determine hematocrit with a centrifuge. We can spin a tube of blood at many times the force of gravity and the solid portions of the blood will settle to the bottom of the tube (figure 16-12). The liquid portion will remain at the top of the tube. Some simple math (divide the height of the solid portion of the sample by the total height of the sample then multiply by 100) provides a percentage that will normally be around 45% for men and about 35% for women. Lower values are called anemia. Higher values are called hyperemia. This simple analysis is a first level doping test for endurance sports. If an athlete's hematocrit is 50% or higher they are generally not allowed to run, bike, or swim. Further testing will be done on an elevated hematocrit sample to determine if the hormone, erythropoietin, was used to artificially increase red blood cell numbers.

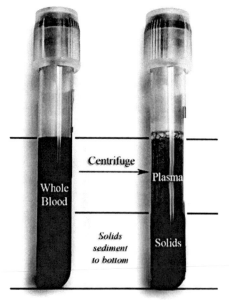

Figure 16-12. Determination of the hematocrit. Measure the height of the solid portion of the blood (right) then divide by the total height of the column of blood (distance between top and bottom lines).

There are two basic types of blood cells, red blood cells and white blood cells (figure 16-13). Red blood cells are more accurately called erythrocytes. White blood cells are more appropriately called leukocytes. Red blood cells primarily carry oxygen to tissues that need them and carry carbon dioxide to the lungs for removal. Higher numbers of erythrocytes (a higher hematocrit) carries with it a functional benefit that can lead to a higher ability to transport and consume oxygen and better endurance performance. As exemplified throughout human anatomy, a simple morphological change produces a profound functional improvement. As such a great deal of interest has been levied at developing exercise protocols increasing hematocrit and unfortunately (on in reference to sport) at developing artificial means of increasing it. There has been success at both. In the latter case, using artificial means to increase hematocrit for sport performance enhancement, there are several instances to hematocrits elevated to the 60s or higher resulting in heart failure during training. Essentially the blood viscosity after drug administration, further increased during exercise, was too thick to pass effectively through small vessels.

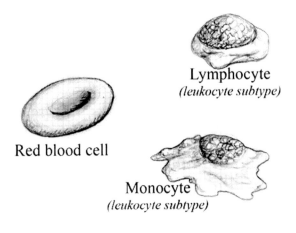

Lymphocyte
(leukocyte subtype)

Red blood cell

Monocyte
(leukocyte subtype)

Figure 16-13. The basic morphology of erythrocytes (red blood cells) and leukocytes (white blood cells).

Leukocytes do not provide exercise performance benefits other than provide infection surveillance and defense services that keep you ready to train. In general, leukocytes produce antibodies or engulf and destroy foreign matter (phagocytosis and lysis). An example of a antibody producing leukocyte is a lymphocyte. An example of a phagocytic cell is a macrophage. Various frequencies and intensities of exercise can produce quite diverse effects of leukocyte numbers. This has been extrapolated to either support or refute the concept that exercise improves disease resistance.

In the early 1900's it was found that hard exercise performed in the early stages of poliovirus infection increased the severity of infection and mortality rate marking the origin of the theory of exercise induced immune suppression. It is not a surprise that attitudes toward engaging in hard exercise were not too positive until after a vaccine for the poliovirus in the 1950's. In fact there are numerous publications from that time referring to severe exercise as harmful. More recent research into exercise induced changes in the blood is quite ambiguous regarding changes in leukocyte morphology.

EXERCISE AND HEART STRUCTURE

The heart is a muscle. It can adapt to being loaded, it can adapt to being unloaded. Where any muscle is concerned, one important adaptation of interest is hypertrophy, the anatomical enlargement of a cell or tissue. Historically hypertrophy of the heart has been considered to be a negative adaptation, a pathology. Regardless of its inducing stimulus, disease or work, it was considered bad and harmful to health. We know better now.

There are three basic types of hypertrophic alterations to cardiac anatomy presented here; pathological hypertrophy, eccentric hypertrophy, and concentric hypertrophy (figure 16-14).

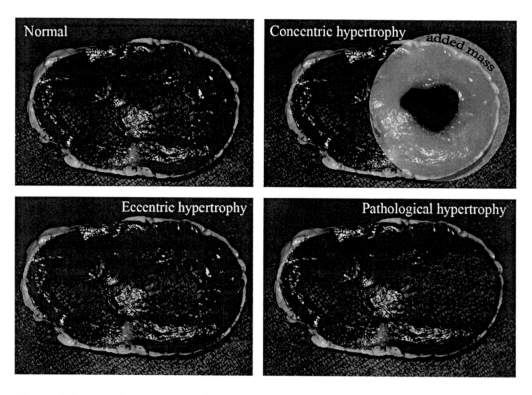

Figure 16-14. Basic types of cardiac hypertrophy. Compare the sizes of the ventricular chambers and the thickness of the ventricular walls.

Pathological hypertrophy can occur as a result of many disease processes such as hypertension, heart attack (myocardial infarction), heart valve disease, and others. The end result is in increase in ventricular mass, potentially up to a 150% increase. While the increased tissue mass is comprised of muscle, it contains significant scar tissue adding no functional capacity. In fact, the existing muscle mass becomes stretched around a larger ventricular chamber, producing thinner free wall exacerbating the hearts already compromised function.

In modern literature, *eccentric hypertrophy* is frequently called, "athlete's heart". As the healthy heart adapts to the increased physical and metabolic demands placed upon it by aerobic exercise (endurance type), the thickness of the free wall minimally increases and the size of the ventricular chambers increase. Combined these two adaptations increase the volume of blood that can be pumped and the force with which it can be pumped. Neither of these adaptations, nor their functional result, can be construed as pathological. Not to say that "athlete's heart" is a misnomer, BUT this is a routine observation in pregnant women, whose hearts adapt and hypertrophy rapidly with the ever-increasing demands placed on their hearts by the developing fetus.

Concentric hypertrophy is another type of athlete's heart where the primary anatomical adaptation to anaerobic exercise (sprint or weight training) is an increase in free wall thickness with little if no change in ventricular chamber dimensions. The functional result of this type of hypertrophy is improvement in the heart's to produce more force and maintain blood flow against resistance levels that previously would have slowed blood delivery. Again, this cannot be construed to be a pathological adaptation.

Although the anatomical and functional adaptations from aerobic and anaerobic exercise do not prevent disease and promote health, they do help the heart function in the face of disease and injury. Think of how beneficial to uninterrupted blood delivery a stronger heart contraction would be if there was a partial occlusion of an artery. The stronger heart just might save a life.

There is a down side of exercise and cardiac muscle anatomy. Luckily it is a side of exercise most recreational trainees will never encounter, severe over training from ultra-endurance training. There has been a number of imaging studies of marathon and ultra-endurance athlete hearts. In those studies it was noted that the cardiac muscle had become scarred, with many lesions present that are not present in normal and healthy hearts. Another study of similar athletes who had trained for marathons for more than a decade showed significant calcification of valvular leaflets, a marker of compromise normally seen in patients diagnosed with cardiac disease. Interpretation of these results is a pesky task, as although the athletes did show pathology similar to cardiac patients, both their cardiac and physical functions were far superior to the cardiac patients. It is suggested that the athlete's may have simply trained more than enough to gain the fitness benefits but exceeded the body's adaptive capacity and caused damage. In other words, they over trained. Unfortunately for the exercising public, and athletes, the existing scientific literature is inadequate to provide relevant training methodologies to improved aerobic fitness without over training. But the emergence of these unusual and pathological findings in endurance athletes should help stimulate adoption of a new perspective on research leading specifically to fact driven recommendations for better exercise programming. The term "run yourself to death" should not be a viable idiom.

VASCULARITY

The last portion of cardiac anatomy we'll consider is actually relevant to the entire cardiovascular system, the basic structure of arteries and veins.

Arteries carry oxygenated blood away from the heart (note the exception mentioned earlier in this chapter). They can be huge, larger in diameter than your thumb, as in the case of the aorta. They can be small arteries, down around 100 millimeters in diameter. Below that diameter the vessels are called arterioles and then terminally on the arterial side we have capillaries, only microns in diameter, big enough to allow passage of only one erythrocyte at a time. The heart generates pressures higher than that in arteries so the blood flows out of the heart into the arteries.

The analog to an artery is a vein. Veins carry deoxygenated blood towards the heart (note the exception mentioned earlier in this chapter). Arteries occur in pairs with veins. A vein that pairs with an artery at any given location is a little larger in diameter than the artery (figure 16-15). An easy example of this (if you pick up a beef heart at the butchers) is the coronary arteries and veins. They are both superficial on the heart surface and easy to see. The larger veins mean they have a lower pressure inside them, allowing the blood from the higher-pressure arteries to flow into them. The right atria has the lowest pressure in it than all of the circulatory elements so the venous blood continues to flow down the pressure gradient into the right atria where it begins the cardiac pressure cycle again.

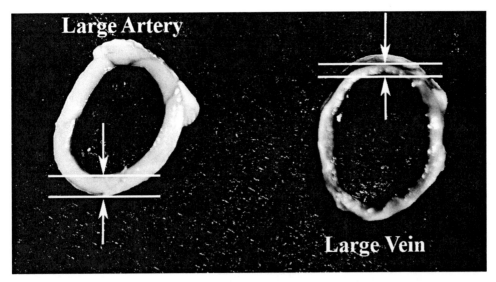

Figure 16-15. Comparison of artery and vein dimensions. Arteries and veins occur as pairs in general. The vein's lumen is larger. The artery has thicker and more rigid walls.

Arteries differ from veins in that the artery is stratified structurally, possesses more of a specific type of muscle – smooth muscle – that aids in directional flow, and has a larger component of an extensible and elastic form of connective tissue in its walls (figure 16-16). Veins are more reminiscent of simple tubes, lacking the more active musculature of arteries. A unique element of veins are the numerous one-way valves that prevent back flow of blood.

Blood vessels are adaptable to stress. If there is a hypoxic stress (low oxygen content present) that results in tissue hypoxemia (low oxygen in a tissue), a cascade of local hormonal and anabolic events occurs that produces new capillaries and new arterioles. This process is called angiogenesis. This is considered to be an endurance friendly anatomical adaptation, leading to an improved oxygen delivery capacity to working skeletal muscle. There is also a survival benefit from angiogenesis that occurs in the heart muscle from exercise too. A more expansive vascular bed from angiogenesis gives the heart muscle more tubes to draw its blood from. If one route of blood delivery is blocked, a newly created collateral vessel or vessels can deliver the needed blood to the heart tissue. Sort of like a traffic detour off of an existing highway after a wreck, where the cars backed up behind the wreck get redirected to other roads but still get to the same intended destination. The creation of collateral blood vessels through exercise driven angiogenesis does not prevent disease processes from occurring. Things like atherosclerotic progression can and do still occur, but angiogenesis can explain, in part, why fit individuals experience myocardial infarctions far less frequently than sedentary individuals.

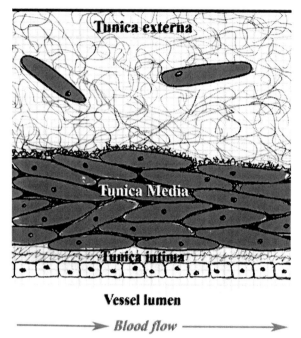

Figure 16-16. The layers of artery and vein structure. The innermost tunica intima is composed of endothelial cells (small cells) and elastic connective tissue elements. The tunica media (middle layer) is comprised mostly of smooth muscle cells (long cells). The tunica externa (exterior layer) is made up of connective tissue with scattered smooth muscle cells. The tunica media and externa are less extensive in veins.

PULMONARY INTEGRATION

The lungs. We can't see them. We can't touch them, but we certainly can feel them when we are really training hard. And they are certainly important to exercise performance in training and in competition. The lungs are intimately intertwined with the heart and vessels in an elegant system of supply and demand of a critical element of life and exercise, oxygen. In most instances when we are hanging around and going about our daily business and even when we are following physicians recommendations for exercise, this elegant system carefully matches the amount of oxygen delivered to amount consumed - supply meets demand. Under these conditions we are using aerobic (oxygen requiring) metabolism to supply the energy need to power these low levels of physical activity. With aerobic exercise, ventilation rate (how many times we breathe per minute) can increase about 300 to 400% during sustained exercise.

If we amp up our exercise intensity - running at near sprint speeds or some similar increase in exertional effort - ventilation rate can exceed baseline by more than 500%. But even that accelerated rate of breathing cannot provide the aerobic metabolic machinery with enough oxygen to keep pace with the demand. As a result there is a transition to anaerobic (non-oxygen requiring) metabolism. Shortly after the transition to anaerobic metabolism, the body will fatigue and exercise will have to slow to an aerobic pace or cease completely. Anaerobic metabolism is a short lived source of power.

So no matter what type of exercise we do, low intensity long slow distance, high intensity sprints, everything in between, and even no exercise at all, the lungs are critical anatomical structures.

LUNG EXTERIOR

Humans have two lungs, each is divided into segments, called lobes that are defined by indentations or creases intruding into the surface of the lung (figure 16-17). The left lung has two lobes and the right lung has three lobes. The lungs sits in the thoracic cavity bounded in front, behind, to the sides, and on top by the axial skeleton. It is bounded below by the diaphragm. As the diaphragm is attached to the contour of the lowest ribs, the lungs can extend no lower. The lungs have a membranous sac surrounding them called the pleural sac. It is a two layered membrane, the outer layer termed the parietal pleura and the inner layer the visceral pleura (figure 16-18). Between the thoracic cavity wall and the parietal pleura is a small amount of pleural fluid that reduces friction between the wall and the pleura during breathing. It also provides a degree of surface tension that keeps the lung surfaces in close proximity to the inner walls of the chest cavity (think about what happens to a piece of plastic wrap if you place it on wet glass, you can slide it around but it is hard to lift off of the glass). Each lung has its own pleural membrane. This is a pretty useful survival asset. If the chest and lung are punctured on one side, a pneumothorax, the other lung is able to function normally, but it will still be extremely uncomfortable.

The lungs have a space in between them, the mediastinum (figure 16-17). Within that space is the heart, thymus, lymph nodes, esophagus, and the trachea, the only atmospheric inlet into the lungs.

Atmospheric air makes its way into the body by first entering through the mouth or nasal passages, passes through the larynx, enters the trachea (windpipe), which then divides into the right and left primary bronchi (plural; bronchi in singular), and then enters the lungs about one third the way down their length - roughly at the mid-sternum level (figure 16-17). The bronchi are cartilaginous (hyaline cartilage) tubes that are also ringed along their length with smooth muscle (the third type of muscle cell).

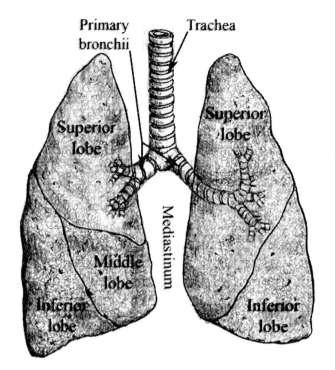

Figure 16-17. Basic parts of the bronchial tree and lungs.

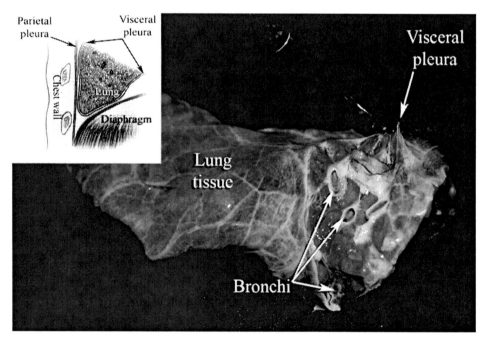

Figure 16-18. The pleura are very thin connective tissue sheets separating the lung from the inner chest wall. The upper left inset shows the spatial relationship of the two pleura. The transverse cut across a lung lobe (photograph) demonstrates how thin the membranes really are.

LUNG INTERIOR

The primary bronchi do not feed into the lungs as most people usually visualize - balloons or big hollow sacs in the chest cavity. This is most definitely not the case. The inside of the lung is filled with millions of membranous sacs, and hundreds of miles of small blood vessels. If you poke a lung, it is not like poking a balloon it is more akin to poking a big, warm, marshmallow.

The lungs are made up of small, spherical air sacs called alveoli. There are about 300 million in a human lung, each about one third of a millimeter in diameter. This provides an extremely large surface area, working out to be approximately the size of a tennis court if spread out into a flat surface. You can calculate this yourself with simple algebra $= 300,000,000(4\pi r^2)$. Pulmonary ("pertaining to the lung") capillaries cover almost the entire alveolar surface and do so in such a complete and thorough fashion that it is almost like a sheet of blood covering the surface of each alveoli. An extremely thin barrier (0.3 μm) lies between these alveoli and pulmonary capillary blood that is known as the "blood-gas barrier". The chemicals present in the barrier allows oxygen to diffuse easily into and carbon dioxide to easily diffuse out of the capillary blood stream. This is important because the heart pushes all of our entire cardiac output through the pulmonary capillaries and past these alveoli where the blood must upload oxygen and offload carbon dioxide.

After a primary bronchus enters a lung, it branches extensively, the tubes getting progressively smaller - macroscopic to microscopic - with every sequential division. There are over one thousand miles of air conducting tubes in the lungs.

The air handling features of the lungs are divided into two zones. the conducting zone and the respiratory zone. The conducting zone is primarily a set of air conducting tubes and gas exchange does not occur between their lumens and the blood. The components of the conducting zone are usually considered to be the trachea, all bronchi, bronchioles, and terminal bronchioles. Think of them as the piping to move air in and out.

The second division, the respiratory zone, is comprised of the respiratory bronchioles, the alveolar ducts, and alveoli (plural, singular is alveolus) (figure 16-19). This is the nuts and bolts of the respiratory tree as all gas exchange occurs in the respiratory zone, with ninety percent occurring at the alveoli. The respiratory bronchioles are the last tube-like structures before the alveolar sacs. The alveolar ducts are short and more irregularly shaped with invaginations along their length. Alveoli are not really perfect little spheres, they are irregularly shaped sacs. They are about one third of a millimeter in diameter on average, but they increase in size with inhalation.

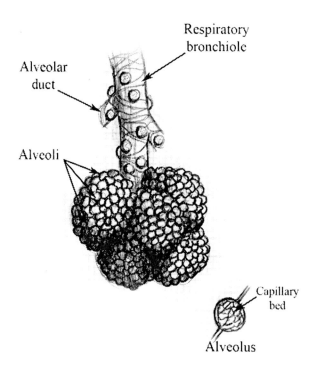

Respiratory bronchiole

Alveolar duct

Alveoli

Capillary bed

Alveolus

Figure 16-19. The respiratory zone of the bronchial tree.

An alveolus is one cell layer thick, made up of epithelial cells. Some of the epithelium cells are simple structural elements. Others are secretory in nature, producing surfactant, a chemical that lowers the resistance to oxygen dissolving from the atmosphere into the blood. It also reduces the stickiness of the inner walls of the alveoli, preventing alveolar collapse.

SUCKING AIR

Asthma, defined as reversible airway obstruction, is a chronic inflammatory disease of the bronchi. During an episode of asthma, an asthma attack, the bronchi become narrowed due to the involuntary contraction of the surrounding smooth muscle. This has been proposed to be a survival mechanism, a reduction in airflow to limit lung injury, that has gone awry in some individuals. A number of external triggers, such as allergy, inhalation of chemical irritants, or exercise, may induce an asthma attack. It has been reported that as many as seven percent of the population has some degree of asthma. An odd observation in regards to sport is that elite athletes seem to have a higher incidence of asthma than the sedentary population, with fifteen percent of the athletes at the 1996 Olympics so diagnosed. There is a possibility that the legitimate frequency of asthma in elite athletes is the same as in normal populations and that the athletes are using the therapeutic drugs as ergogenic aids - as they dilate the bronchi and reduce respiratory work (improved breathing efficiency). As the highest incidence of athlete diagnosis is in endurance sports where respiratory efficiency is imperative for elite success, this suggestion may have some validity. However, asthma can be deadly, with about four thousand deaths in the USA annually, and banning therapeutic drugs is likely not a viable course.

The prevalence of asthma in humans has increased over the past few decades. The reasons proposed for this phenomenon is myriad, but one of the most unique proposals is that today's children are brought up in too aseptic of an environment, they aren't vaccinated against environmental allergens through normal and habitual exposure as their parents were. Regardless of the reason for the increase in prevalence, coaches and fitness trainers will work with asthmatic athletes and must become cognizant of how to safeguard their trainees in the case of an episode.

"In summary, a zombie outbreak is likely to lead to the collapse of civilisation, unless it is dealt with quickly. While aggressive quarantine may contain the epidemic, or a cure may lead to coexistence of humans and zombies, the most effective way to contain the rise of the undead is to hit hard and hit often."

– Munz, Hudea, Imad & Smith
in Infectious Disease Modelling Research Progress (2009)

17 – BALANCE & COORDINATION

Skeleton, muscle, heart and lungs. Everything you need to move, right? No, human movement is an excruciatingly complex result of numerous environmental and internal inputs and countless internal processes that all culminate to produce carefully titrated force and magnitude of movement - in a tiny area of the body, of the body as a whole, or of the body and external objects as a system.

One contributing system that has to be considered in any examination of human movement is the nervous system, a network of specialized cells called neurons that coordinate, propagate, transmit, and coordinate electrical communication within the body. The human nervous system is divided into two basic anatomical groups, the central and peripheral nervous systems. You can draw a loose parallel to the axial and appendicular skeleton here with the central nervous system consisting of the brain, retina of the eyes, and spinal cord. The peripheral nervous system would be everything else - nerves and the neurons (nerve cells) that comprise them.

NEURONS & NERVES

A neuron is the basic unit of the nervous system, an electrically responsive cell that is capable of transmitting electrical information to other cells. Neurons have characteristic anatomical features, a soma or cell body, dendrites - a collection of filaments arranged in a spreading branch orientation, and an axon that connects the soma to another neuron's dendritic tree (figure 17-1). Axons can end as a single terminus or branch and have multiple axonal termini. Axons can also be myelinated (layered with lipid) or be un-myelinated. Myelination aids in speed of neural transmission.

Neurons are the longest cells of the body, some stretching for as long as a meter (39 inches). A group of axons collected together into varied size bundles and travel to various areas of the body are called nerves. The dendritic termini can be spread across and are on the order of many hundreds of microns (micrometers). Where one neuron communicates with another cell there is a microscopic space between the two, a synaptic cleft or junction. There are chemicals released into these spaces, neurotransmitters, that aid the electrical signal to cross or to "synapse". These signals move along the axon of one neuron to a dendrite or cell body of another (there are exceptions to this general template). There are on the order of one hundred billion neurons networked across the human body sharing information.

In the most simplistic analysis of nerves, there are two different types of nerves, sensory nerves and motor nerves. Sensory nerves receive chemical and physical stimuli and covert them into an organized electrical signal that can be transmitted along its length. Sensory nerves are activated by chemical and physical stimuli impinging on them, producing a non-homeostatic cellular environment and signals that travel to the central nervous system, providing information relative to the status of specific cells, organs, systems, and the environment. Motor nerves receive information from sensory or other neural components and convert the neural signals into a stimulus that drives contraction of muscles or glandular activity.

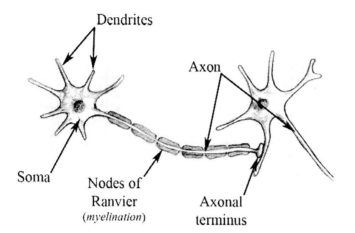

Figure 17-1. Basic features of a neuron and its synapse with a second neuron.

Once a neural signal reaches its target cell or tissue, it does not have to cause an excitatory response, it can cause excitation, inhibition, or some variant in between the two.

SENSORY RECEPTORS

There a several types of sensory neurons relevant to human movement, those that act like tactile sensors at the skin and those that act as mechanical receptors within the muscles.

Pacinian corpuscles - These are nerve endings embedded in the skin. They produce the sensation of changing pressure and are also nociceptors, responsible for sensitivity to pain. A Pacinian corpuscle is about one millimeter in length. When considered in a three dimensional view it is somewhat similar to a oval pancake. There is an inner cavity filled with fluid and a single nerve ending. The cavity is wrapped in twenty or more thin layers of connective tissue, each separated from the next by a viscous gel (figure 17-2). When the layers (lamellae) of the corpuscle are deformed from pressure applied or from pressure

288

an electrical signal is created by the movement of sodium ions. If the stimulus is large enough, the signal will be transmitted up the neural tree. This means that the response here is a graded one, enabling identification of degrees of pressure applied. Large pressures, or rapidly applied pressures, make the neuron send out neural signals faster than smaller or more slowly applied pressures. This mechanism is called rate encoding. Out in the real world, this means that these nerves are stimulated to fire when the skin gets deformed by pressure quickly. If the pressure is applied very slowly and continuously, there is no generated signal. The Pascinian corpuscle's also become deactivated fairly rapidly. Once a pressure is applied and the initial neural response and feedback are complete, they "get used to" the pressure, and the distortion of the lamellae becomes the referent baseline, thus no further signals are generated or sent. Think about the sensations on the skin when you first put clothes on. After a few minutes, you no longer "feel" their presence, a result of a mobile homeostasis in these cells. This is fairly useful for human perception as it eliminates background neural clutter.

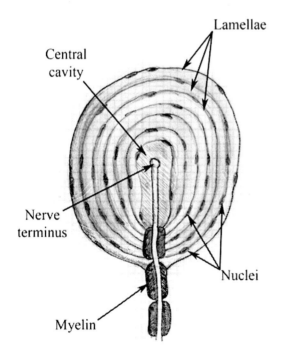

Figure 17-2. Schematic of a Pascinian corpuscle.

Meissner's corpuscles - These nerve endings are also embedded in the skin, primarily at the fingers, lips, and aereoli, and produce sensitivity to light touch. The corpuscles are flattened layer structures, with a single nerve ending and a thin surrounding connective tissue capsular membrane (figure 17-3). They are much smaller than the Pascinian corpuscles, about one tenth the size. Any physical impingement and deformation, no matter how slight, will produce an electrical stimulus to be produced. They are particularly affected by touch and vibration. These would be the sensory tools with which you detect a mosquito landing on your skin. Like the Pacinian corpuscles, these receptors adapt to the new pressure stimulus if it is constant, creating a new homeostatic condition.

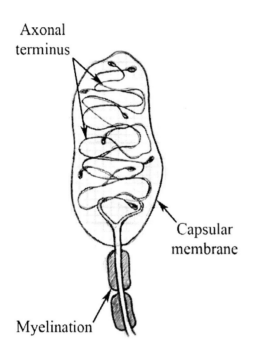

Figure 17-3. Schematic of a Meissner's corpuscle.

Merkel nerve endings - These receptors are again found in the skin, but can also be found in the mucosal membranes. They are very sensitive and provide detailed touch information, specifically regarding pressure and texture, to the central nervous system. Each receptor possesses a Merkel cell that is in close proximity to a nerve terminus (figure 17-4). They are found in superficial skin layers such as just beneath the ridges of the fingerprints and in certain sensitive areas of hairy skin. These receptors do not respond to rapid changes in pressure they are more persistent in nature and respond to slowly impinging stimuli. They also respond differentially to shape. Sharp elicits a more powerful response than dull. And convex is less powerful than concave.

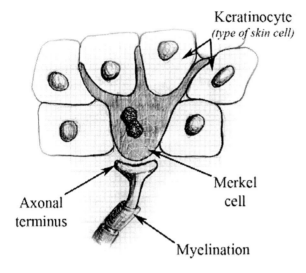

Keratinocyte
(type of skin cell)

Figure 17-4. Diagram of a Merkel cell in relation to an axonal terminus and the skin cells from which it receives stimuli.

Merkel cell

Axonal terminus

Myelination

Ruffini corpuscles - This receptor is a spindle-shaped (cigar shaped) structure located in the skin. As with the other receptors of the skin, a thin connective tissue sheath covers a web of dendritic termini (figure 17-5). These receptors are responsive to stretching forces along the skin - think about how the skin bends and stretches during movement, that distortion is what gets sensed - and aids in detecting changes in joint angle. These receptors are very dense in the hands and help control finger position and movement. They also detect slippage of objects as they begin to slide along the surface of the skin, this allows reaction and adjustments in grip position and strength.

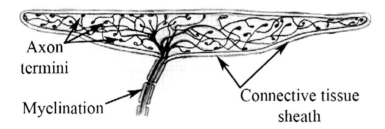

Axon termini

Myelination

Connective tissue sheath

Figure 17-5. A Ruffini ending of the skin.

Interfusal fibers - Interfusal fibers, also called muscle spindles after their shape and location, are found lying parallel to extrafusal fibers (regular muscle fibers) in the belly of a muscle. The spindle apparatus contains three or more small intrafusal muscle fibers, spiraling dendrites surround the fibers, and a thin connective tissue covering (figure 17-6). These receptors are stretch receptors, detecting changes in muscle length. Muscle length information can be relayed to the central nervous system via sensory neurons for higher level processing to aid in determining the position of the body and its component parts in space. Length information can also be relayed to the spine via sensory neurons then back to the originating muscle through motor neurons, this would be an example of a spinal reflex, an involuntary muscle action directly responding to sensory input. In this reflex, interfusal fibers regulate the contraction of a muscle to resist muscle stretch, the *stretch reflex*. Essentially the sensory neuron components of the receptor that wrap around the modified muscle cells of the spindle are separated by an impinging stretch. This deformation induces an electrical impulse to be sent up the neural chain. At the spinal cord, the impulse is redirected down a motor neuron back to the muscle experiencing the stretch. The stimulus from the motor neuron causes a muscle contraction in the opposite direction of the stretch. This is an oft referred to and desirable reflex in sport. If the "stretch reflex" can be invoked at a precisely timed moment, the force of a muscle contraction increases. An example of this would be a high force or high velocity eccentric movement followed immediately by a high force or high velocity movement in the opposite direction. Slow movements or low velocity loaded movements do not produce a stretch reflex. Further, static stretching (all variants) when held for a brief time provides a transient homeostatic state where the reflex is silent. An example of how powerful this reflex is can be found in a simple experiment. Squat down six inches and hold it for ten seconds, then jump (do not succumb to the temptation to squat down further before you start the jump). Touch as high as you can on a wall. Now stand erect, squat down six inches quickly and immediately jump and touch as high on the wall as you can. You will note a much higher jump with the quick dip and drive method, the stretch reflex in action.

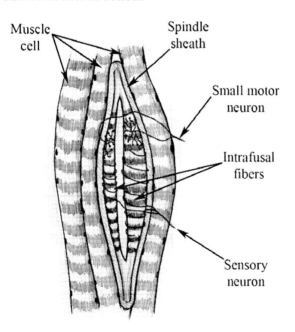

Figure 17-6. Basic structure of intrafusal fibers and their innervation.

Golgi tendon organs - The Golgi tendon organ, commonly known simply by the acronym GTO, is a force sensing receptor located in the muscle at the area where the muscle cells phase out and the coalescence of the endomysium, perimysium, and epimysium into tendons. The receptor is composed of a sensory nerve with its dendritic endings coiling around collagen strands of the tendon (figure 17-7). Whereas muscle spindles are organized longitudingally along the muscle to sense stretch, the Golgi tendon organ is arranged to sense stimuli laterally across the tendon. When a muscle contracts, its tendon is compressed, squeezing the terminal ends of the dendrites. This deformation produces a neural stimulus that is passed up the neural chain. Each Golgi tendon organ monitors about ten to twenty motor units. One of the major attributed functions of this receptor is an inhibition reflex, as forces increase, the signal produced results in a reduction in force produced. It is frequently held that this receptor acts in this manner as a protective mechanism to prevent injury when a high force load is applied rapidly - if you can curl seventy five pounds but someone places two hundred and fifty pounds in your hands (arms bent at ninety degrees) you will immediately drop the weight reflexively as soon as the magnitude of the load is sensed as injurious. The golgi tendon organ may contribute to this reflex but it is active in providing sensory information throughout the complete range of forces experienced by the muscle, not just high force. A second major function of the receptor is in excitation of muscle during walking and running. In this role it provides feedback that results in contractile activity that aids in the execution and timing of the load-bearing and non-load-bearing phases of gait.

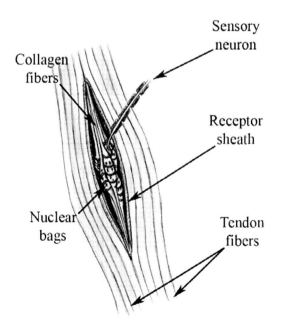

Figure 17-7. The golgi tendon organ.

Whether is tossing a ball with precision towards a target or putting a weighted ball for distance, the neuroreceptors assist in correctly titrating the amount of force required and in calculating the trajectory that will most likely result in success.

BRAIN

The brain is the processing center for much of the information produced by the sensory receptors. It interprets input and produces appropriate motor output, consciously or unconsciously. A complex structure comprised of about thirty billion cells, the various regions of the brain have a diverse set of functional responsibility, but yet act in a most cohesive manner (figure 17-8). Some of the regions are important to sport and exercise.

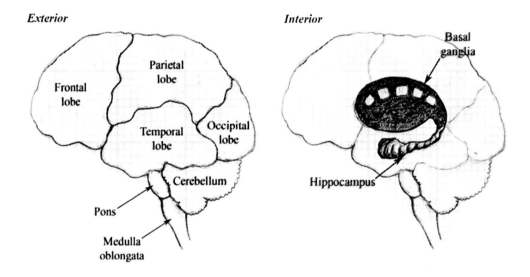

Figure 17-8. Lateral schematic of the regions of the brain.

Medulla oblongata - This appears as an extension of the spinal cord (it really isn't), the portion that widens just inferior to the main mass of the brain. It receives a wide variety of sensory input and controls a number of involuntary (autonomic) functions. Of relevance to exercise is its control of respiratory rate, blood pressure, and cardiac function. Peripherally related to exercise is its control of the reflexes of swallowing and vomiting (coughing and sneezing too).

Cerebellum - The cerebellum is the center for coordination, precision, and synchrony (timing) of movement. Input from sensory neurons, along with information from other parts of the brain are modified here in order to refine motor output. Of extreme importance to exercise and sport is that this is the structure that allows us to learn, by trial and error, how to perform a specific motor task. When we first get on a bike to learn how to ride, we are wobbly and erratic. But as we repeat the activity, the cerebellum makes adjustments that narrows the gap between actual activity patterns and the intended activity patterns. Without the cerebellum we could never learn new sports or move in a smooth and fluid manner.

Hippocampus - This small region of the brain is related to exercise only in that it functions in memory and in awareness of ones location in the environment (sort of "where you are on the map").

Basal ganglia - The basal ganglia are a set of small structures on the interior of the forebrain. They have a primary function in selecting appropriate motor actions. Action selection is choosing the next appropriate behavior to produce, whether it is motor or cognitive.

Cerebral cortex - The cerebral cortex is a convoluted sheet of neural tissue, the outermost layer of the brain (figure 17-9). It carries out a wide variety of functions influencing memory, attention, perceptual awareness, thought, language, and movement. It is your gray matter. The cortex possesses both sensory and motor areas. The sensory areas are the visual, auditory and somatosensory cortexes. Input to these areas comes from the opposite side of the body - input from the right ear goes to the left auditory cortex for example. The motor area of the brain is the motor cortex. Sub-areas of the motor cortex either select which voluntary movement to perform or executes a motor pattern for a voluntary movement.

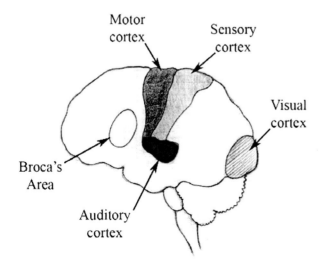

Figure 17-9. Lateral view of the major areas of the cerebral cortex.

CONTROLLING NEURAL ACTIVITY

Exercise and movements specific to sport are motor skills, learned sequences of movements that, when combined appropriately, produce a fluid and efficient execution of the particular exercise or sport task. There are two basic types of motor skills, gross (large scale) and fine (small scale). Gross motor skills are things like walking, jumping, squatting, throwing, lifting, and many more. Fine motor skills are much more related to hand-eye coordination activities like manipulating small objects or may involve the use of very precise motor small-scale movements. The minute hand movements when throwing a curve ball, or when aiming a target pistol would be examples. But all movement is not either or fine or gross motor activity, there is a continuum and it may not be clear cut. Lifting is a gross motor activity, but in the snatch - an Olympic weightlifting event - deviance in bar path by as little as a couple centimeters can lead to technical failure, an example of a gross motor activity with a fine motor component.

Control of gross and fine motor activities is accomplished in the brain, it computes the solutions to motor problems. If an exercise task requires the shoulder angle to be x and the elbow angle to be y at time z, the brain will select a course of action to arrive at that spatial and temporal orientation as efficiently as possible within its innate and experiential domain. A reflex, or innate movement will be simple and efficient, it is the most direct means of achieving the movement goal. Conversely, we learn many movements through trial and error. It is proposed that approximately three hundred to eight hundred movements in the same motor pattern will ingrain that pattern into the brain's repertoire of movement choices. If the movement was not learned correctly by the trainee or not taught correctly by the trainer, then the engrained pattern will be inefficient. This can be corrected by overwhelming repetition of desired, correct, and efficient motor technique.

EARS

Effective movement within any environment requires sensory input from the ears. The obvious auditory (sound) cues we are familiar with - when you hear a blitzing linebacker closing from your blind side for example - are created in the cochlea in the inner ear (figure 17-11). There is also another part of the ear that is essential to balance and determining our body orientation in space - it tells us that we are face down on the turf after being pancaked by the blitzing linebacker - the vestibular system of the inner ear (figure 17-12).

Figure 17-10. Fine motor activity can be something like hitting a target with a tennis ball (left). Gross motor activity can be something like throwing a medicine ball for distance (right).

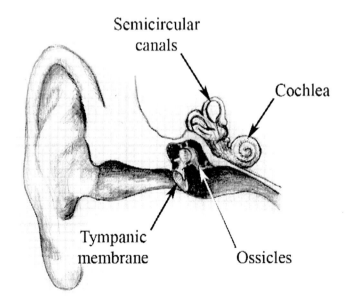

Figure 17-11. Features of the inner ear.

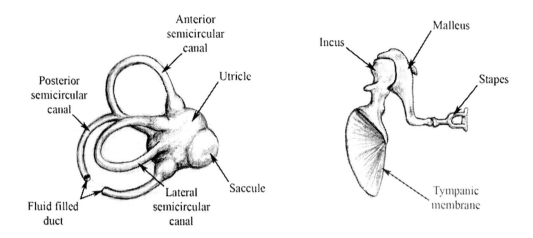

Figure 17-12. The semicircular canals (left) and the ossicles (right).

In the middle ear, the sound (pressure) waves produced by some environmental entity is converted into mechanical vibrations at the tympanic membrane (ear drum). There are small bones, or ossicles, on the inside of the membrane that articulate with the cochlea that receive these mechanical signals and introduce them to the fluid inside the cochlea. The three bones are called the malleus (or hammer), incus (or anvil) and the stapes (stirrup)(figure 17-12).

Inside the cochlea there are hundreds of thousands of tiny hair-like receptors that convert the wave energy into neural impulses which are transmitted to the brain for interpretation. The vestibular system contains the semicircular canals, a set of three hollow ring-like structures oriented at slight angles to the three cardinal axes (frontal, transverse, and sagittal). Each canal is lined with sensory hairs that produce information sent to the brain about the tilt, rotation, and linear motion of the head. The brain then interprets the information to provide motor output to maintain balance during even the most complex of motion tasks. As the head moves along an axis, the sensitive hairs lining the canal oriented along the axis of movement are bent and send information to the brain. Slowing or stopping the movement allows the return of the hairs to straight(er) and another positional data point is relayed to the brain. An interesting and fun pass-time for some is to overwhelm the positional detection capacity of the vestibular system by putting ones head on a bat and spinning for a minute then trying to run in a straight line. When the hilarity ensues - running sideways and falling down - it is because the head down position on the bat accompanied by the rapid rotations of the head around the axis sensed by the superior canal put a strong current into that canal. When the head position is raised, the brain is still sensing a persisting rotational flow and cannot detect the new orientation until the flow stills, thus disorientation occurs for a few moments.

In many instances, when the signals sent to the brain by the semicircular canals do not match what the eyes see, nausea and vomiting may result. This would be the case when the canals sense the pitch and yaw of a boat but the eyes see a stable constant image inside the cabin - you get seasick.

EYES

Eyes are organs that use their generous compliment of receptors to detect light and convert it to neural impulses that can be sent to the brain for interpretation and action. The eyes are often imagined as round balls, that's why we call them eyeballs. But they are not truly round, they conform to the shape of the orbital socket in the skull, they are ball-like. If the eye is unusually shaped, corrective glasses are usually required. The eye is a hollow three layer structure. The outermost layer is the cornea and sclera. The middle layer has the choroid, ciliary body (where the lens is), and the iris (where your eye color is - the pupil is the black dot in the middle of it). The inner layer is the retina, the sensory part of the system. The space inside the eye is called the vitreous humor and is filled with transparent gel (figure 17-13). Light passing through the lens, through the vitreous humour, impinges on the rods and cones, and stimulates production of neural impulse that are relayed to the brain.

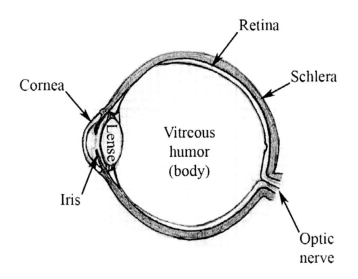

Figure 17-13. A cross sectional diagram of the eye (transverse section).

The eye can detect about sixteen million different colors and has a dynamic contrast ratio (the same scale of light to dark as used in rating HD televisions) of 1,000,000:1. Colors are detected by a specific photoreceptor type in the retina, cones. Cones are most active in high light conditions. Rods, provide black and white discrimination and are most active in low light conditions. There are about ten rods for every cone making this, day-night, functional division viable.

The eye functions as a joint, it has seven muscles crossing the articulation of the eye and the skull; the levator palpebrae superioris, lateral rectus, medial rectus, inferior rectus, superior rectus, inferior oblique, and the superior oblique (figure 17-14). All rotate the eye within its orbital socket. Each muscle is effuse with motor neurons making extremely small and controlled movements possible.

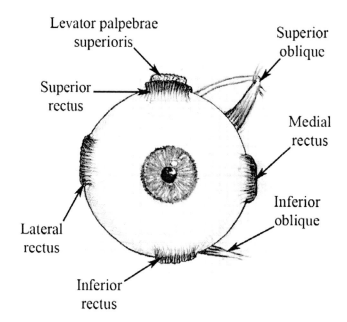

Figure 17-14. The muscles of the eye.

HAND-EYE COORDINATION

The eyes are an essential component of movement in exercise, sport, or otherwise. Without visual feedback we would not be able to easily locate the glass on the table to pick it up, would be unable to efficiently target its top in order to fill it, we would not be able to readily detect when it was full, nor could we gracefully sit the glass down on the table. Imagine your sport played or loading a bar in the gym without the advantage of sight - humbling.

Control of hand-eye coordination involves virtually every part of the nervous system. The peripheral sensory receptors, the eyes, the spinal cord, the brain, all play an important part in organizing and sequencing the movements of the eyes and the hands. Hand-eye coordination has been studied for at least a century as it is essential in combat and work. Research related specifically to sport is much more recent. It needs to be understood that all the previous research is very applicable to exercise and sport as the mechanisms of control are identical. While the eye is integral in hand-eye coordination, the eye does play a similar role to everything else-to-eye coordination. From selecting foot position before commencing an exercise to moving the whole body in response to seeing a particular movement in an opposing player, the importance of vision cannot be understated.

"The inner contents of my mind are an enigma."

– Patrick Starfish

18 – FAT

Human survival over the eons has hinged on the ability to survive periods of food scarcity - winters, droughts etc. As a consequence the ability of humans to store metabolic energy in the body ensured that the species would survive from season to season (in pre-agrarian times). If you take a look back into the pteroglyphs and the oldest surviving representations of the human form you will note that in most instances a very rounded (what we would call obese looking) figure is represented. The Willendorf Venus (Naturhistorisches Museum, Vienna) from approximately 20,000 bc presents a human female form in quite a corpulent form, sort of a bowling ball shape with appendages and parts attached. A more recent discovery in Germany's Hohle Fels cave (University of Tuebingen, Germany) dating to at least 33,000 bc also shows a very full-figured female form. There is much debate about motivations and meanings of such ancient discoveries, but it may just be as simple as that a person with extensive fat deposits made for a better mate in those stark times as they were more likely to survive over time than a skinny counterpart. These early efforts at representing the world around the artist may not have been sculptures of real people, rather sculptures representing valued characteristics at the time. There is no evidence that would suggest a school of abstract art was operating during paleolithic times so we have to assume that the figures are idealizations based on observed successful biological traits. As distasteful as some might find this, like so many other animals, our ancestors might have been carrying around a bunch of extra pounds by then end of the growing season when food was plentiful. They also were likely rather wispy by the end of the winters seasons where food availability was limited.

So it seems as though humans have developed to be able to gain fat and lose fat - gain it during times of nutrient surplus and lose it during times of nutrient deficit.

More recently, no one really considered fat as unhealthy and no one considered fat particularly healthy for many centuries. Rather is was often considered a status symbol or visual expression of wealth. Where do you think the saying "fat cat" came from? Only the wealthy could afford the luxury of excess food and a lifestyle with no physical labor that led to a fat body. This perception of fat as a positive persisted in American culture into the early 1900's and vestiges can still be seen in some Polynesian cultures today.

Here is where the modern problem with obesity begins. The Paleolithic fat man was only seasonally fat and never was sedentary. The privileged Victorian man and woman were fat and sedentary rear round. When mechanization and technology reduced the amount of labor required for virtually all occupations, obesity and sedentarism could be enjoyed by all. This is quite a divergence from our original anatomical design and functions. Fat and active versus fat and sedentary. The new combination is not necessarily a good one.

Research often tries to correlate high body fat to early death and you will see quite a bit of such literature in journals and sensationalized in the media. But there is no single damning piece of data linking fat to death. As late as the 1990's obesity was not even considered a primary risk factor in the development of cardiovascular diseases. It is only when we consider obese individuals with sedentary lifestyles and other risk elevating behaviors that the relationship becomes very strong. It is lack of exercise that is problematic, not fatness per se. When we gave up the Paleolithic habit of activity, we set the stage for early death. Several large scale studies have clearly demonstrated that being strong leads to a longer life span. Others have demonstrated that higher levels of endurance are associated with longer life spans. This sets up a nice little stratification of survivability:

> Low to moderate body fat + strength + endurance = Best combination for longevity
> Low to moderate body fat + strength = Next best combination
> Low to moderate body fat + endurance = Next best combination
> Low to moderate body fat + sedentary = Unacceptable

If we change the fat storage amounts a bit we come up with another stratification:

> High body fat + strength + endurance = Good combination for longevity
> High body fat + strength = Next best combination
> High body fat + endurance = Acceptable combination
> High body fat and inactive = Unacceptable

Such stratifications are somewhat useful as they give us an idea of what we need to do in terms of improving longevity (but there are always exceptions and life provides us no guarantees). But all the preceding does not touch on the anatomy of what many consider the enemy, body fat.

Inside metabolically active cells you will find little droplets of fat (more properly called lipid) scattered about that can be used as an energy source. It should be understood that fat carries out many other crucial roles in the body, besides being a pretty effective energy storage medium. At the cellular level, every one of the billions of cells comprising our body requires fat as a component of its membrane structures. At the biochemical level, a tremendous number of important chemicals in the body are composed of or are created from fats, estrogen and testosterone for example. Neuronal axons are myelinated (chemical made from fat) in order to make neural transmission effective. We have to have fat in our bodies for normal function and in our diet to replace the fats used and degraded every minute of every day. Fat is everywhere.

But these are not the fatty deposits that we are so visually familiar with and have an almost cult like obsession with their elimination. At the gross level, conglomerations of fat containing cells, adipocytes, form padding and shock absorption structures surrounding the heart and other visceral organs. In its subcutaneous form adipose tissue (accumulations of adipocytes) provides for a long term energy source, a layer of insulation from both thermal challenges, and also protects against physical insult too.

How much energy is stored in the body's fat deposits? If we consider a 100 kilogram pound person (220 pounds) with 25% body fat, there is 25 kilograms of fat present (55 pounds). If each gram of fat (1000 grams per kilogram) represents 9 kcal (calories) of stored energy, then there are 225,000 stored calories. This quantity of energy dwarfs the body's stores of its other main energy source, carbohydrate, with about 600 grams or 2400 calories of available carbohydrate stores in muscle and the liver (about 1% of muscle mass and 10% of liver mass). It would take about 90 days to deplete that much fat if we could mobilize and consume 2500 of those calories each day, could obtain other required nutritional elements in a calorie free form, and a bunch of other minor things that make a no food intake scenario problematic. It would take just a few hours of moderately intense, continuous, and prolonged exercise to deplete carbohydrate stores (carbohydrate depletion is one of the contributors to distance athletes "hitting the wall").

MICROANATOMY

If you look at fat cells under a microscope you will note that they look very reminiscent of the arrangement of styrofoam beads in the wall of a cooler. White fat cells, also known as monovacuolar cells, are essentially large lipid droplets surrounded by a thin layer of cytoplasm. The lipids stored inside the cell are primarily triglycerides (three fatty acids bound to a glycerol backbone). The cells are mononucleate with the nuclei found on the periphery of the cell. There is a huge range in cell size seen in humans. Although the "average" size is frequently reported to be 0.1 millimeters, these cells can be much much smaller or tremendously larger depending on the nutritional state of the individual. Fatter individuals will tend to have larger, hypertrophied fat cells compared to their slimmer counterparts. In this manner, fat cells and fat tissue are adaptive entities that can be affected by nutritional habit and by exercise regimen. Fat cells can adapt to store more energy by simply adding more fat to its contents, but there is a critical volume, when about four times the normal content of fat is present, the the adipocyte will mitotically divide to open up new storage space.

This occurs in extreme obesity and most of us are not in any danger of adding more and new fat cells. What we are in danger of is adding new triglycerides to the interior of our existing adipocytes simply by eating a caloric excess or by under-exerting ourselves to burn too few calories.

Normal
adipocyte

Hypertrophied
adipocyte

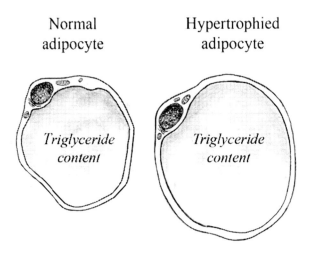

Figure 18-1. Adipose cells gain mass by adding lipid content to the interior of the cell. Essentially the lipids contained form a large droplet suspended in the small amount of cytoplasm present.

In infants and to a much lesser extent adults there is a second type of fat tissue, brown fat. Composed of a subtype of adipocyte that is more metabolically active than white fat, brown fat produces heat that helps keep infants warm. These cells are multinucleate, have more cytoplasm, have more bioenergetic organelles, and less fat content than white fat cells. In adults brown fat is essentially absent, found in exceedingly small amounts in the neck and shoulders in some humans.

Although adipocytes comprise the major mass of cells present in adipose tissue, fibroblasts, macrophages (leukocyte found in tissue not blood), and endothelial cells are all present as part of the structural unit. As one would expect, there are quite a few blood vessels perfusing fat tissue. This facilitates fat deposition (adding fat to a adipocyte) and fat mobilization (removing fat from an adipocyte).

ANATOMICAL DISTRIBUTION

Adipose cell accumulations, adipose tissue, are located subcutaneously (under the skin), viscerally (surrounding internal organs), in bone marrow, and in breast tissue. There is a great deal of variation of morphology within the subcutaneous and visceral deposits between individuals and between sexes. Both the shape of deposit and location of deposit seem to be affected by male and female sex hormones (testosterone and estrogen). Males tend to accumulate more abdominal visceral fat than females and females tend to accumulate more lower body subcutaneous fat than males. No matter the location, fat deposits are not smooth sheets of tissue. They are opportunistic in where the develop, conforming themselves to any irregular nook or cranny available. As such they can be fingerlike projections, amorphous blobs, or anything in between.

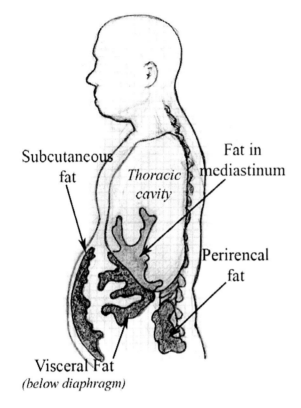

Figure 18-2. Common deposition locations for fat.

Subcutaneous fat

Thoracic cavity

Fat in mediastinum

Perirencal fat

Visceral Fat
(below diaphragm)

Subcutaneous fat - The largest repository of fat is found under the skin. If you pinch the skin on the back of your hand you are basically pinching the epidermis and dermis, as there is very little subcutaneous fat on your hand. Half of the thickness pinched is the thickness of those two layers. The thickness of the epidermis and dermis vary by anatomical location by anatomical region, but the back of the hand assessment gives us a starting point to look at how thick the layer of subcutaneous fat can be. Pick anywhere else on the body and pinch. The difference between that pinch and the pinch on the back of the hand is primarily fat. That's the foundation of the skin fold measurement technique of determining body fat, the thicker the fold, the more body fat present.

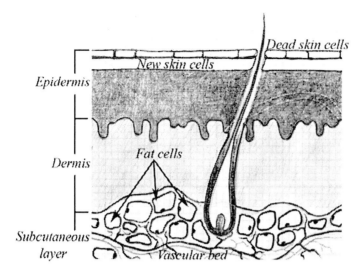

Epidermis

Dead skin cells

New skin cells

Dermis

Fat cells

Subcutaneous
layer

Vascular bed

Figure 18-3.
Subcutaneous fat
deposits are superficial
and can easily be
palpated.

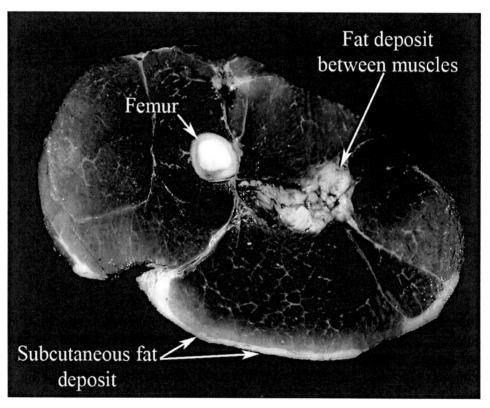

Fat deposit
between muscles

Femur

Subcutaneous fat
deposit

Figure 18-4. It's easy to get a feel for what and where fat is found in the body, just look at what is displayed in the grocery store meat isle. A ham is shown here, pork chops are from along the thoracic spine, and bacon (think of how much fat is in the slices you cook) is the abdominal wall.

There is a frequent misconception that there are different types of subcutaneous fat (this is not in reference to white v. brown fat). The basic concept is that there is regular fat and then there is cellulite, a reportedly different type of fat responsive only to special methods of fat reduction. There is no such anatomical distinction. The term cellulite was coined by a French cosmetologist in the last century ... coined as a means to market beauty products. "Cellulite" is simply fat cell clusters that have hypertrophied sufficiently to press against the skin to produce an irregular surface appearance. Nothing more, nothing less.

Visceral fat - Also known as abdominal fat, these are the deep deposits found around the internal organs of the peritoneal or abdominal cavity. There are specific regions of visceral fat named according to location; mesenteric (near the gastrointestinal system), epididymal (near the reproductive system), and perirenal (near the kidneys).

Although the cell structure of visceral fat is the same as that of subcutaneous fat, it is harder to lose as it is replenished as fast or some cases faster than it is depleted.

FAT DEPOSITION AND DISEASE

It is an extremely common practice in the fitness community to measure subcutaneous fat thickness, use a formula to express the results as a percent of body mass, and call it acceptably healthy if it is lower than 25% in males and 30% in females. Anything over that level would be considered contributory to cardiovascular diseases, diabetes, and a variety of other pathologies. This is not a new concept, but the amount of fat deemed acceptable is under constant revision. In the 1960's Kellog's popularized a concept of "if you can pinch an inch" you need to diet by eating Special K cereal. Recently a new picture is emerging, visceral fat is presently being indicted as the "bad" fat, more related to cardiovascular disease than subcutaneous fat. You can't pinch visceral fat. Further, the presence of subcutaneous body fat has been proposed as a protective agent against such diseases in individuals with high levels of visceral fat.

It is likely that there will be debate on how much fat is good, how much fat is bad, where are the good deposits, where are the bad ones, and such for decades to come. Research looking at correlations, as virtually all human obesity studies do, will generally not answer questions about causality.

In sport and fitness practice, the rule of thumb is that moderation is key. Body fat percentages of 5% or less in males or 12% or less in females is too low to support hard training and performance, and likely are representative of an inappropriate diet or over training. Percentages in the 20s and 30s are too high.

Athletes need to be somewhere in the middle. In sports where power-to-weight ratios are important to success, like cycling, a low percentage nearing the lower limit is acceptable and desirable. In weight class sports where head-to-head combat or competition are the norm, a low percentage is also an advantage. In virtually all other types of sport, a moderate level of adipocity is just fine and pushing the lower limits is unnecessary. It can still be pursued if desired, a male football athlete at 11% can drop to 5%, but it accomplishes very little performance benefit. In fact, a cautionary note is that dietary restriction to lose body fat will at the same time reduce the anabolic capacity of the body as a whole. Post-training recovery will be slower and muscle development capacity will be less, meaning progress will come at a slower rate. Balancing fat loss with training goals is a razors edge to walk.

INDEX

"And so life returned to normal, well, as normal as life gets on this planet populated by psychotic apes."

- Nibbler

Professor Lon Kilgore has been teaching fitness physiology and exercise anatomy in undergraduate pre-physical therapy curriculum and graduate exercise physiology programs for nearly two decades. He has developed a unique perspective and approach relative to the application of science to sport and exercise that he passes on to his students - or anyone else who will listen. He graduated from Lincoln University with a Bachelor of Science in Biology, earned a Masters in Kinesiology, and a Ph.D. in Anatomy and Physiology from Kansas State University. He has competed in weightlifting to the national level since 1972 and coached his first athletes to national championship event medals in 1974. He has also competed(s) in powerlifting, wrestling, rowing, and golf. Thirty nine years after he started training, he still sets lifetime PRs at least once each year. He has worked in-the-trenches as a coach, as a sports science consultant with athletes from rank novices to professionals and the Olympic elite, and as a head university strength coach. His interest in developing better weightlifting coaches, strength coaches, and fitness professionals have driven much of his academic and professional efforts. He spent a decade as a certifying instructor for USA Weightlifting and was a frequent lecturer and researcher at the US Olympic Training Center in Colorado Springs. His authorship efforts include books, magazine columns, and research journal publications. His illustration efforts have similarly appeared in many books, journals, and online.

"Sometimes it takes a crazy person to see the truth.
If so, I'm a freaking lunatic."

- Stephen Colbert

CPSIA information can be obtained at www.ICGtesting.com
Printed in the USA
LVOW100031261011

252036LV00017B/45/P